Acceptance and Diffusion of Connected and Automated Driving in Japan and Germany

Christine Eisenmann · Dennis Seibert ·
Torsten Fleischer · Ayako Taniguchi ·
Takashi Oguchi
Editors

Acceptance and Diffusion of Connected and Automated Driving in Japan and Germany

 Springer

Editors
Christine Eisenmann
Institute of Transport Research
German Aerospace Center DLR
Berlin, Germany

Dennis Seibert
Institute of Transport Research
German Aerospace Center DLR
Berlin, Germany

Torsten Fleischer
ITAS
Karlsruhe Institute of Technology
Karlsruhe, Baden-Württemberg, Germany

Ayako Taniguchi
Department of Risk Engineering
University of Tsukuba
Tsukuba, Japan

Takashi Oguchi
Institute of Industrial Science
University of Tokyo
Tokyo, Japan

ISBN 978-3-031-59875-3   ISBN 978-3-031-59876-0 (eBook)
https://doi.org/10.1007/978-3-031-59876-0

This Springer imprint is published by the registered company Springer Nature Switzerland AG
The registered company address is: Gewerbestrasse 11, 6330 Cham, Switzerland

If disposing of this product, please recycle the paper.

Contents

Introduction

Christine Eisenmann, Dennis Seibert, and Christian Winkler

Abstract This opening chapter provides an initial overview of the various topics covered in the book, which explores CAD's implications, focusing on societal acceptance and technology diffusion. Key areas include the impact of existing mobility systems on CAD adoption, policy frameworks, ride-hailing market potential, and public perception in Germany and Japan. Through modeling and empirical studies, the research examines CAD's effects on transport systems, car ownership, and travel demand. This Japanese-German collaboration highlights CAD's importance for both countries, given their advanced automotive industries and demographic challenges, offering insights for future research and policy development.

Connected and Automated Driving (CAD) technologies are expected to increasingly penetrate road transport in the next decades. CAD consists of two aspects: the first—"connected"—means vehicles communicate with each other (Vehicle-to-Vehicle or V2V communication), or with infrastructure (Vehicle-to-infrastructure, or V2I communication), such as traffic lights [3]. The second aspect—"automated"—is related to driving systems which—depending on automation level (see below)—take over some or all driving tasks. Vehicle connectivity and vehicle automation can develop independently. However, the combination contributes to unfolding the potential of this new technology in terms of safety, efficiency and comfort [2]. The combination of the two terms, and consequently the two technologies, leads to the designation "connected and automated driving", or CAD, and implies utilizing the benefits of putting these technologies together. Figure 1 illustrates the SAE Levels of Driving Automation that we refer to in this book as the automation levels.

The expected benefits of CAD are increased safety, smoother traffic flow [1, 5], and reduced congestion, with the possibility for drivers to engage in other activities en route, and to maintain accessibility in regions with low demand for public transport services, or ensuring mobility for people with mobility constraints [4, 5]. CAD has the

C. Eisenmann (✉) · D. Seibert · C. Winkler
German Aerospace Center (DLR), Institute of Transport Research, Berlin, Germany

D. Seibert
e-mail: dennis.seibert@dlr.de

C. Eisenmann et al. (eds.), *Acceptance and Diffusion of Connected and Automated Driving in Japan and Germany*, https://doi.org/10.1007/978-3-031-59876-0_1

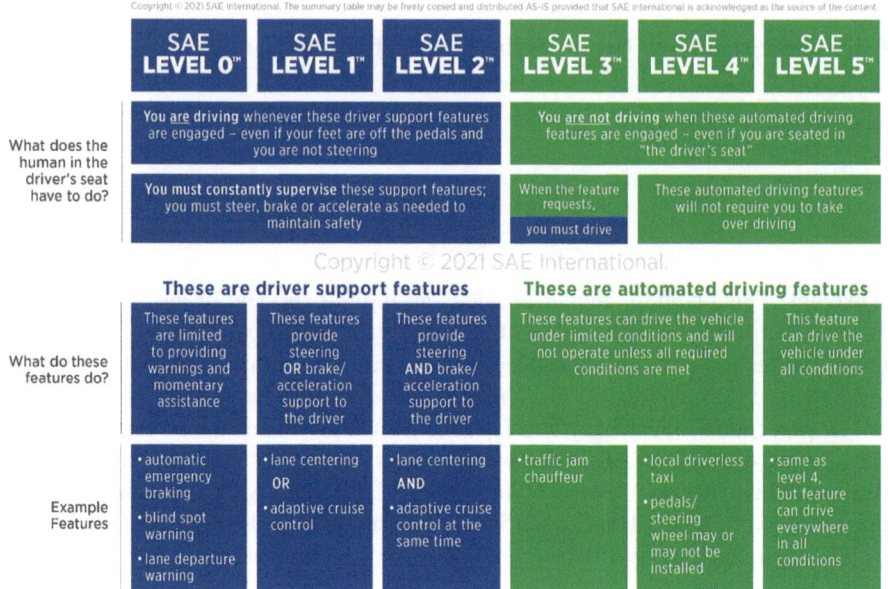

Fig. 1 SAE J3016 levels of driving automation. *Source* © SAE International from SAE J3016™ Taxonomy and Definitions for Terms Related to Driving Automation Systems for On-Road Motor Vehicles (2021–04–30), https://saemobilus.sae.org/content/J3016_202104/

potential to change multiple dimensions of the transport system, ranging from altered patterns of car ownership, through the rise of new mobility services, to changes in traffic flow or even spatial structures. Far reaching socioeconomic impacts appear likely: Industrial production and value chains in the automotive and transport sectors may be affected substantially through the advent of CAD, not least because business models might change as new mobility services and players enter the scene.

The wide deployment of CAD depends on various factors: most significant are the rate of adoption and the rate of diffusion of CAD vehicles and services, as well as changes in individual travel patterns and collective travel demand that result from the availability and use of CAD technologies. The Japanese-German research collaboration which has resulted in this book has therefore addressed two closely interrelated issues of impact assessment, in order to: (1) understand factors that contribute to the perception, adoption or rejection of CAD, phenomena which are usually framed as part of "societal acceptance" in public and political discourses, and (2) understand, describe and model the diffusion of CAD technology.

The book addresses a range of aspects along the two overarching themes of **societal acceptance** and **technology diffusion**:

- Today, **framework conditions for travel** shape our mobility and travel behavior. It is to be expected that the existing mobility systems are key as to how CAD will be adopted in different geographic, demographic, cultural and economic contexts. A comprehensive analysis of existing framework conditions of the prevailing mobility systems in Germany and Japan and their likely impact on the adoption of CAD is presented in **chapter** "Setting the Scene for Automated Mobility: A Comparative Introduction to the Mobility Systems in Germany and Japan".

- One specific framework condition that relates to the development and deployment of automated driving technologies and services in both Japan and Germany are policy processes. **Chapter** "Governance, Policy and Regulation in the Field of Automated Driving: A focus on Japan and Germany" investigates in particular how the governance style relates to **regulatory changes and resource allocation** in the development of new technologies and innovations in the two societies.

- The **ride-hailing market** is expected to grow significantly alongside the advancements in automated driving technology. However, profitability analyses are scarce, owing to insufficient data relating to underlying cost structures. A holistic business analysis and prognosis of the ride-hailing market is, therefore, highly beneficial in the current discussion and is presented for Germany in **chapter** "Business Analysis and Prognosis Regarding the Shared Autonomous Vehicle Market in Germany".

- Research and policy agree that **acceptance of CAD** is key to its diffusion. Therefore, defining social acceptance and elaborating the relationships between the subjects and objects of acceptance in the context of automated driving is crucial to understand the full scope of the concept. **Chapter** "Social Acceptance of CAD in Japan and Germany: Conceptual Issues and Empirical Insights" presents this work, enriched by empirical insights from original quantitative surveys performed within the research collaboration. Additionally, the chapter discusses changes in attitudes towards CAD and explores how automated vehicles are covered in newspapers.

- It is undisputed that the **diffusion of CAD** will have major impacts on the transport system. **Chapter** "Transportation Effects of Connected and Automated Driving in Germany" investigates the effects of CAD on various parameters of the **German transport system** by means of three consecutive transport models. Among these parameters are the development of car ownership, the diffusion of automated vehicles in the passenger car fleet, and multiple transport demand variables. Modeling insights suggest that CAD per se will not be accompanied by favorable outcomes for the transport system.

- Transportation effects of CAD depend on the framework conditions of the prevailing mobility systems. Moreover, the effects of CAD may differ when applying different modelling approaches. **Chapter** "Transportation Effects of CAD in Japan" addresses the effects of the **diffusion of CAD** on the **Japanese transport system** by means of a modeling approach that differs from that used in **chapter** "Transportation Effects of Connected and Automated Driving in Germany". Modeling insights partially point in a similar direction to the results from the German modeling, but the Japanese analysis shows that enhancement of consumer expectations will be crucial for ensuring the spread of automated vehicles.

- **Chapter** "Overall Comparison between Germany and Japan in Relation to Social Impact of Connected and Automated Driving" concludes with a summary of the research undertaken in chapters "Setting the Scene for Automated Mobility: A Comparative Introduction to the Mobility Systems in Germany and Japan–Transportation Effects of CAD in Japan", and outlines implications for future research about CAD and its role in an efficient and sustainable mobility system.

The present book is a result of the Japanese-German Research Co-operation on Connected and Automated Driving. Research on CAD is of high relevance for both countries due to several reasons. Both countries are characterized of a crucial role of the automotive industry for past and future growth and employment. Moreover, they represent prototypes of mature mobility markets that many emerging economies are likely to follow in the next decades. Japan is characterized by a mix of high urban densities with low car use and mountainous areas with a small population with high car use while Germany features medium density, medium sized cities and average European car use. Demographic change is ubiquitous in both countries implying challenges for current and future mobility systems.

On the German side, research was conducted within the CADIA project (Connected and Automated Driving: Impact Assessment, 2019–22) funded by the Federal Ministry of Education and Research. The three-year project was led by the Institute of Transport Research at the German Aerospace Center (DLR). Project partners were the Institute for Technology Assessment and System Analysis (ITAS) at the Karlsruhe Institute of Technology (KIT), the Chair and Institute of Urban and Transport Planning at the RWTH Aachen University, and the Bayerische Motoren Werke AG (BMW).

On the Japanese side, two research groups were involved in the collaboration. A collaborative research group from the University of Tokyo and Doshisha University on '*Study of Socioeconomic Impacts of Automated Driving Including Traffic Accident Reduction*' (JPNP18012) was commissioned by the New Energy and Industrial Technology Development Organization (NEDO), under the National R&D Program *SIP-adus* (Automated Driving for Universal Services, Strategic Innovation Promotion program) led by Japanese Cabinet Office (CAO). The research group led by Prof. Ayako Taniguchi at the University of Tsukuba, conducting empirical studies on social acceptance of automated driving from various angles, was supported by the Japan Society for the Promotion of Science (JSPS) KAKENHI Grant Number 20K20491.

Researchers from both countries developed a deep understanding of the two transport systems and the potential impact of CAD on these systems. Through close contact by means of virtual and physical meetings, conferences, and workshops ideas, the approaches, and results of the research conducted were discussed transparently. Meetings in Berlin and Kyoto contributed significantly to a better understanding of the research, but also to cultural and personal exchanges. This book represents the result of close cooperation between Germany and Japan. It shall contribute to the further joint research activities in the spirit of the close partnership between the two countries.

References

1. Beiker, S. A. (2015). Einführungsszenarien für höhergradig automatisierte Straßenfahrzeuge. In M. Maurer, C. J. Gerdes, B. Lenz, & H. Winner (Eds.), *Autonomes Fahren: Technische, rechtliche und gesellschaftliche Aspekte* (197–217). Springer.
2. e-mobil BW GmbH-Landesagentur für Elektromobilität und Brennstoffzellentechnologie, Prognos AG, Fraunhofer-Institut für Verkehrs- und Infrastruktursysteme IVI, TÜV Rheinland Consulting GmbH, & Technische Universität Berlin. (2015). *Automatisiert. Vernetzt. Elektrisch.: Potenziale innovativer Mobilitätslösungen für Baden-Württemberg.*
3. Gerpott, T. J. (2021). Connected Car. In T. Kollmann (Ed.), *Handbuch Digitale Wirtschaft* (pp. 1071–1089). Springer Gabler.
4. Lenz, B., & Fraedrich, E. (2015). Neue Mobilitätskonzepte und autonomes Fahren: Potenziale der Veränderung. In M. Maurer, C. J. Gerdes, B. Lenz, & H. Winner (Eds.), *Autonomes Fahren: Technische, rechtliche und gesellschaftliche Aspekte* (pp. 175–195). Springer.
5. Schreurs, M. A., & Steuwer, S. D. (2015). Autonomous Driving - Political, Legal, Social, and Sustainability Dimensions. In M. Maurer, C. J. Gerdes, B. Lenz, & H. Winner (Eds.), *Autonomes Fahren: Technische, rechtliche und gesellschaftliche Aspekte* (pp. 151–173). Springer.

Setting the Scene for Automated Mobility: A Comparative Introduction to the Mobility Systems in Germany and Japan

Tobias Kuhnimhof, Hiroaki Miyoshi, Ayako Taniguchi, Shoichi Suzuki, Yu Hasegawa, Torsten Fleischer, and Christine Eisenmann

Abstract This chapter presents a comparative analysis of the mobility systems in Germany and Japan, providing insights into how these systems might influence the implementation of vehicle automation. This comparison begins by exploring the historical evolution of transport in both countries, noting that both have long-established infrastructures shaped by unique geographical and historical contexts. Germany's transport system, for instance, developed within a landlocked nation with extensive rail networks, while Japan's transport was influenced by its island geography and mountainous terrain. The chapter then examines key dimensions

The original version of the chapter has been revised: The author's corrections have been updated. A correction to this chapter can be found at
https://doi.org/10.1007/978-3-031-59876-0_9

T. Kuhnimhof (✉)
Institute of Transport Planning, RWTH Aachen University, Aachen, Germany
e-mail: kuhnimhof@isb.rwth-aachen.de

H. Miyoshi
Faculty of Policy Studies, Doshisha University, Kyoto, Japan
e-mail: hmiyoshi@mail.doshisha.ac.jp

A. Taniguchi
Faculty of Engineering, Information and Systems, University of Tsukuba, Tsukuba, Japan
e-mail: taniguchi@risk.tsukuba.ac.jp

S. Suzuki · Y. Hasegawa
Institute of Industrial Science, The University of Tokyo, Tokyo, Japan
e-mail: suzuki41@iis.u-tokyo.ac.jp

Y. Hasegawa
e-mail: yuhase@iis.u-tokyo.ac.jp

T. Fleischer
Institute for Technology Assessment (ITAS), Karlsruhe Institute of Technology (KIT), Karlsruhe, Germany
e-mail: torsten.fleischer@kit.edu

C. Eisenmann
Institute of Transport Research, German Aerospace Center (DLR), Berlin, Germany

of the current transport systems, including demography, settlement patterns, road transport governance, public transport infrastructure, and the automotive industry's role. Comparative statistics are provided, illustrating the differences and similarities between Germany and Japan. The analysis highlights how these existing systems serve as both enablers and barriers to the integration of automated vehicles. The chapter concludes that the introduction of vehicle automation will not revolutionize these transport systems overnight but will gradually adapt to existing frameworks. The success of vehicle automation depends on the interplay between technological advances and established transport policies, regulations, and cultural norms. This chapter suggests that understanding the deep-rooted structures of transport systems in Germany and Japan can offer valuable insights into how vehicle automation might unfold in other regions with mature mobility markets. In conclusion, the chapter provides a holistic framework for analyzing the potential impacts of vehicle automation, stressing the importance of considering the existing transport system's legacy and the multifaceted nature of mobility in Germany and Japan.

1 Introduction

Vehicle automation is being discussed as a possible game-changer with the potential to revolutionize transportation. For about a decade, automation has been one of the key topics of transport research, and there have been visions of imminent disruptive changes to the transport system caused by automation [3, 61, 70]. However, automation technologies in transport are not going to be introduced into undeveloped transport systems. Instead, they will be implemented in mature mobility markets with established framework conditions, ranging from infrastructural conditions, through legal foundations for transport, to deeply ingrained mobility routines on the side of travelers [8]. In other words: Implementation of automated vehicles (or AVs) will not all of a sudden establish an entirely new transport system in which they unfold their full potential. Instead vehicle automation technology will have to adapt to existing framework conditions, and most likely successively alter these conditions in order to unfold its potential step by step. In order to understand this process and allow for projections regarding the stepwise implementation and effects of vehicle automation, it is key to understand the pillars that define today's transport system, and their relationship to vehicle automation.

Japan and Germany are predestined mobility markets which are useful case studies to understand the relevant domains of existing transport systems and their consequences for the implementation of vehicle automation. Germany represents a European transport system that has taken a middle way between the strongly car-oriented North American evolution of the transport system, and the less car-oriented eastern Asian experience, for which Japan represents a prototypical path [30, 68]. Assuming that the path of extensive car orientation that North America has taken is unsustainable for many emerging economies, it appears likely that the future of emerging mobility markets in the world falls between the European and the Japanese approaches [107].

Hence, studying the Japanese and German transport systems and the implications for the implementation of vehicle automation will allow for conclusions as to what to expect in other transport systems in the future.

This chapter takes a different approach to the rest of this book, and thus provides both a multifaceted prelude and a complementary methodology. The other chapters mostly focus on specific dimensions of vehicle automation, and often take a quantitative approach. In contrast, this chapter looks at various dimensions of the transport system in a cross-national comparison, and draws multidimensional conceptual conclusions.

After introducing the historic evolution of the transport systems in Germany and Japan, this chapter selects ten topics that are key to how these systems are constituted today. Each topic section presents comparative information for the two countries, and some include comparative statistics. Each section concludes with an interpretation of how these pillars of the transport system have influenced or will influence the implementation of vehicle automation. These discussions of the dimensions of the transport systems in Germany and Japan provide a holistic framework for speculation to identify drivers and barriers of vehicle automation, and how it will change established transport systems in the decades to come.

2 Historic Evolution of Today's Transport System

2.1 Germany

Elements of today's transport system in Germany date back to ancient times, such as Roman or medieval urban street layouts in some city centers [48]. However, much of the foundation of the modern German transport system was laid in the 19th century with the advent of the railway. Plans for long distance rail networks preceded the opening of the first German rail line in 1835 [125]. Specifically, between 1840 and 1880, the German long-distance rail network expanded rapidly, reaching its largest extent early in the 20th century. About 90% of today's long-distance rail lines in Germany originate from that era. From the second half of the 19th century, urban rail networks extended quickly [46]. At first, these were horse-drawn, but from the turn of the 20th century electric catenary tramways took over [47]. This development coincided with rapid population growth, specifically in German cities, which were characterized by extremely high population densities relative to today's standards. However, the increased travel speeds brought about by urban tramways for the first time in history allowed cities to grow substantially in spatial size while keeping total travel times in check. While in the early days long distance and urban rail development was mostly financed and operated by private entrepreneurs, gradually the state took over from the end of the 19th century [47]. There was a mix of motivations to do so, ranging from military strategy considerations to the prospect of generating revenue from railway operations. However, from that time onward, German transport planning and policy was characterized by the notion that the transport system could not be left to market forces but required strong state intervention.

The early 20th century saw the emergence of the automobile as a new mode of transport. At first, automobiles represented a risky leisure activity for tech-savvy and adventurous young, and mostly male, urban elites [69]. While Germany, like other European countries, lagged behind the United States with regard to motorization by several decades, cars started to become more commonplace from the 1920s [68]. Much of the motor vehicle regulation which is in place today in Germany originated in this era, including requirements to register and insure vehicles, and rules regarding driver education and vehicle taxation. Due to effective automobilist lobbying, in the early days vehicle tax revenue was hypothecated for road construction, providing substantial funding for roads and fueling road construction from the 1920s [47]. Contrary to popular belief, the first federal Highway ("Autobahn") was not inaugurated by the Nazi Regime, but in 1932 by the then mayor of Cologne, Konrad Adenauer (who became the first German Chancellor in 1949). After the Nazis came to power in 1933, they put a propaganda focus on automobiles and continued extension of the Autobahn construction. Contrary to their propaganda, however, automobiles continued to be a status symbol of elites, while the train remained the dominant mode of transport for the general population [47].

After World War II (1939–45) the growth of car ownership and use in Germany increased such that total automobile travel surpassed total travel by public transport (1955) and cars outnumbered motorcycles (1957) [19]. This development went along with strong economic growth, auto-oriented post-war reconstruction of German cities according to the principle of separating land uses, and a surge in suburbanization. Specifically, in the 1960s, transport policy and planning were strongly car-oriented [106]. Many cities replaced the tramways with bus systems and underground urban rail to make space for urban roads [46]. In the 1960s, there were a few years with hypothecation of fuel tax income to, initially, road construction, and then transport infrastructure construction in general. This provided generous funding and caused today's German cities to be partly characterized by over-dimensioned road and transport infrastructure.

From the 1970s onward, transport policy and planning paradigms began to change, as there was increasing awareness of topics such as energy dependence (specifically oil), environmental damage, or quality of life in cities. The foundation of the German Green Party in 1980 epitomizes this change in perspective. Specifically, in urban transport this paradigm shift manifested itself in an increased focus on public transport, a partial push-back of the car in central urban areas (traffic calming, reduction of public parking spaces etc.) and—from the 1990s—more focus on cycling infrastructure planning [106]. Nevertheless, the regulatory heritage (e.g., urban parking statutes, road transport law) or established planning conventions (e.g., for highway infrastructure) leads to an underlying structure for transport policy and planning that remains relatively car-oriented.

2.2 Japan

The main transportation routes in Japan are influenced by the ancient road network "Shichidou Ekiro", developed in the 7th century [89]. The Shichidou Ekiro was made up of seven wide, straight arterial roads, constructed by the Imperial Court to connect the central government and local counties. There was also a post-horse transportation system called "Ekiden-sei", similar to the Cursus Publicus developed by the Roman Empire [118]. However, pedestrian, horse, and boat traffic persisted until the Meiji Restoration in 1868, because wheel traffic did not develop in Japan.

After the Meiji Restoration, Japan's first railway was constructed in 1872 between Shinbashi and Yokohama. From this point, railway development progressed rapidly. For road traffic, horse-drawn stagecoaches commenced operating in 1868 between Yokohama and Odawara, followed by carriage operation between Tokyo and Takasaki in 1872 [105]. However, railroad construction was prioritized, and road construction did not progress. After the Sino-Japanese War (1894–1895), the railway network expanded rapidly, and in the 1910s, a comprehensive domestic transportation network was formed by the railway and coastal shipping networks.

Automobile importation began in 1901 in Japan; however, in 1913, the number of cars in use was only approximately 500, whereas more than 170,000 horse-drawn wagons were responsible for freight transportation. With modernization of the economic structure, the demand for road infrastructure development has increased. The Road Law was enacted in 1918, which included rules regarding road types, grades, management responsibilities, and cost burden. Japan's first long-term plan for road development was established in 1919; this comprised a 30-year planning period starting from 1920 [72]. However, owing to the economic recession after World War I (1914–18), and the Great Kanto Earthquake in 1923, the budget was reduced, and road development did not proceed. During WW II, resources were used for the war effort, and the level of road infrastructure remained low.

The Road Law was amended in 1952, and the National Expressway Law was enacted in 1957, resulting in the addition of the National Expressway to the road types. In terms of financial resources and planning, the first five-year road development plan was formulated in 1954, and road construction and improvement using the fuel tax as a specific financial resource began in earnest. In 1956, the Road Development Special Measures Law was enacted, and the Japan Highway Public Corporation was established to launch the existing toll road system.

In March 1947, the number of registered automobiles was more than 140,000 which increased significantly to approximately 77 million by March 2015. In addition, the length of national highways has increased approximately six times during the 66 years from 1949 to 2015 [56]. In 2005, the Japan Highway Public Corporation was privatized, and in 2006, the construction and operation of the expressway based on this privatization scheme commenced.

In recent years, road development and management has been pursued not only for automobile traffic but from other viewpoints. In 2016, the Bicycle Utilization

Promotion Law was enacted, and the basic measures promoting the utilization of bicycles were comprehensively clarified. In 2020, "The 2040 Vision for Roads in Japan [88]" was published by the Ministry of Land, Infrastructure, Transport and Tourism (MLIT). It states that road policy sets human wellbeing at the center, and addresses issues and needs such as climate change, aging infrastructure, declining population, digital transformation, and new post-COVID-19 lifestyles, by using digital technology and restoring the function of roads as a livable communication space.

2.3 Comparative Conclusions

The early historic evolution of the transport systems in Germany and Japan was formed by different political conditions and was specifically shaped by fundamentally different physical preconditions: Large parts of Germany are landlocked, and as a result the country's transport infrastructure is characterized by intensive land transport interaction with its neighboring countries. Japan consists of islands which—in addition—are largely characterized by a challenging mountainous topography. In both countries, the transport systems underwent fundamental changes in the last 150 years, specifically with the advent of railways and automobiles. Hence, disruptions and revolutions in the transport sector are not new to Germany or Japan. However, in the case of both railways and automobiles it took decades until the new technologies unfolded their potential and reached their peak. This is partly due to the fact that there is always a multidimensional transport system heritage, ranging from physical infrastructures, through the legal framework, to societal, cultural and psychological settings. Vehicle automation may represent a similar revolution in the transport sector in the long-term. However, the speed with which this transition takes place likely depends on the number of required changes in the existing transport system heritage. While road vehicle automation may not require much adaptation in the physical infrastructure (which is still unclear), there are obviously fundamental changes as regards other dimensions of the transport system (e.g., the regulatory framework or digital infrastructure). The historic review of technological transitions in transport in Germany and Japan suggests that it may well take many decades to overcome the inertia of the transport system.

Table 1 Key figures on demography and settlement patters [36, 37, 71, 114]

	Germany	Japan
Total population	83 million (2020)	126 million (2020)
Proportion of population under 20	18% (2020)	16% (2020)
Proportion of population 20 to under 60	53% (2020)	49% (2020)
Proportion of population 60 and over	29% (2020)	35% (2020)
Proportion of population in cities under 100.000 population	68% (2020)	29% (2020)
Proportion of population in cities with 100.000 to under 1.000.000 population	22% (2020)	47% (2020)
Proportion of population in cities with 1.000.000 population and more	10% (2020)	24% (2020)
Population density (persons/hectare) in urbanized areas	54 (2013)	67* (2020)
Number of municipalities	10,787 (2021)	1,741 (2021)

* Population density of Japan is that of the densely inhabited districts

3 Demography and Settlement Patterns

3.1 Comparative Statistics

3.2 Germany

After centuries of high birth and death rates, as in other European countries, death rates in Germany declined substantially from the second half of the 19th century, while birth rates continued to be high [13]. Hence, this period was characterized by strong population growth superimposed with urbanization due to industrialization in urban areas. As a consequence, most German cities grew unprecedentedly during this era, for example Berlin by factor 10 [49], Aachen by factor 5 [110], and Essen grew by a factor of 100, partly due to incorporation of neighboring villages [127]. Despite such discrepancies in urbanization, growth was not concentrated in a few metropolises but spread across many cities, and the polycentric spatial structure that characterizes Germany today emerged. Many German cities took on more or less their current population size between 1850 and 1950. Thereafter they have mostly expanded in terms of space rather than population.

After WWII and following an influx of about 12 million post-war refugees, in 1950 the German population (west plus east) was about 70 million and continued to grow for another two decades [10]. This was mostly due to an annual birth surplus of about 400,000 more births than deaths. Germany reached a population of about 80 million around 1970—approximately the population it still has in 2022. However, birth rates started to drop drastically in the 1960s, and since around 1970 the relation of births and deaths has reversed, with a continued surplus of deaths over births of up to 200.000 in Germany per year. Since then, shrinkage of the German population was offset by immigration, which took place in several waves. The first waves (1960 and 1970s) mostly included migrants from southern Europe and the Mediterranean. Later waves were fueled by migrants from Eastern Europe and the former Soviet Union (1990s) and other EU countries, as well as the Middle East and Africa (after 2010). Today, about a quarter of the German population has a family background of international migration [10].

As immigration was mostly concentrated in urban and economically well-off rural areas, some rural areas—mostly, but not only, in eastern Germany—are severely affected by aging and shrinking of the population. For Germany as a whole this is not apparent yet, but this is likely to change as the 'baby boomer' generation (born between 1955 and 1969)—which currently still makes up a large part of the workforce—will retire in the next decade.

The aging of the population and an increasing proportion of senior single- or two-person households also contributes to increased consumption of living space per capita in Germany, which has grown from 33 m^2 per person to over 45 m^2 per person between 1990 and 2020 [38]. But declining household sizes and growth of living space per capita is not only a phenomenon caused by aging; it prevails in all age groups. This has contributed to declining population densities in built-up urban areas in Germany in recent decades, despite urban planning efforts to increase urban densities in order to curb land consumption for settlement. Overall, today Germany has average European population densities which continue to decline, even as metropolitan areas grow in population because of continued sprawl and suburbanization [71].

3.3 Japan

Japan had a population of 34.8 million in 1872, at the beginning of the Meiji era, but this began to grow rapidly, reaching 51.3 million in 1913, at the beginning of the Taisho era, 72.1 million by the end of WWII in 1945, and peaking at 128.1 million in 2008 [116]. Since then, Japan's population has been plummeting, and—using medium variant projections—is predicted to decrease to 83.2 million by 2070 and to 50.56 million by 2115 [95]. Japan is thus poised to experience a phenomenon without precedent in the history of human civilization: a population that, after a century of growth, undergoes a century of decline and then returns to where it began [80].

Transformative changes are apparent not only in Japan's total population, but also in the age distribution of its citizens. In 1970, at the heart of Japan's rapid-growth era, the proportions of the population represented by young people (below 15), working-age people (15–64), and older people (65 and above) were 23.9%, 69.0%, and 7.1% respectively, but by 2020 these numbers had shifted to 11.9, 59.5, and 28.6% [114]. The prediction for 2050 is 10.6, 51.8, and 37.7% [95]. This rapid depopulation and aging, unprecedented in world history, is a natural consequence of two trends: the rapid demographic transition Japan experienced as a late-emerging developed nation, and a total fertility rate that, for many years, has fallen well below the population-replacement level. Looking at households, as of 2020 Japan had 54.1 million households, of which 35.7% were one-person households and 55.9% were nuclear families; by 2040, the number of households is expected to fall to 50.8 million, of which some 40% will be one-person households—with senior citizens accounting for more than half [96]. Thus "parents-and-children" households, traditionally the most common variety, are becoming a minority, and will soon be outnumbered by senior citizens living alone.

An aging society with declining birth rates will soon find itself short of workers, and around the time of the bubble economy in the late 1980s, Japan attempted to compensate for this shortage by accepting foreign laborers. At present, the number of foreign nationals in Japan is estimated to be 2.7 million; 2.2% of the population in 2020 [114]. In Japan, "immigrants" are distinguished from "foreign workers". Foreign nationals working in Japan are accepted only for limited periods of time; unlike the U.S., Canada, and Australia, Japan has no system for admitting foreign nationals or their families as permanent residents (that is, immigrants) in a systematic way each year.

As for the geographical distribution of the Japanese population, after WWII, residents of rural Japan flocked to cities to study and work; the rural populations of Eastern and Western Japan were absorbed primarily into the Tokyo, Nagoya and Osaka areas, respectively. In Europe, foreign workers supported post-war growth, but in Japan it was this internal population migration that supported the growth of cities.

A characteristic feature of Japan's geographical population distribution is the prominent concentration of the population in cities. Of Japan's 47 prefectures, Tokyo is the most densely populated, with 6,410.4 persons per km^2 in 2020—some 19 times the national average of 338.2 persons/km^2. Tokyo is followed by Osaka and 5 other prefectures with populations exceeding 1,000 persons/km^2. At the other extreme, there are four prefectures with fewer than 100 persons/km^2: Kochi, Akita, Iwate, and Hokkaido, whose population densities are 1/66, 1/78, 1/81, and 1/96 that of Tokyo [114]. Incidentally, we note that the 23 wards of Tokyo, which had received a net annual influx of residents for many years, crossed-over to a net outflux in 2021 [115]. This, however, did not reverse the overall trend of increasing concentration of population in the Tokyo metropolitan area and is just one indication of a trend—driven by the COVID-19 pandemic and the rise of remote working—that may shape the distribution of Japan's population in coming years.

When comparing municipal populations and population densities in Japan and Germany, the substantially different total number of municipalities (*Shi-Chō-Son* in Japan; *Gemeinde* in Germany) must be taken into account (Table 1). Both countries have undergone a long history of reducing the number of municipalities, with the objective of providing basic services such as education, police, administration etc. effectively and efficiently [78, 108]. This, however, has led to far fewer municipalities in Japan relative to Germany, which are also about ten times larger than their German counterparts [60]. As a result, Japanese municipalities are often much more heterogeneous with regard to land use, population densities and transportation network characteristics.

3.4 Comparative Conclusions

Current and future demographic trends in both countries seem to align well with vehicle automation. Firstly, there is the aging and long-term shrinkage of the population in Japan, which is a global forerunner when it comes to such demographic change, and in Germany, which is among the most rapidly aging countries in Europe. As a consequence of aging, an imminent lack of workforce is inevitable, and replacing human drivers through automation technologies is an obvious approach to address this challenge. Secondly, aging populations are characterized by an increasing proportion of travelers with mobility impairments, even if these may occur at a later age in the future. Travelers with mobility impairments may be among those who benefit the most from automation technologies if they are relieved from the task of operating vehicles. Thirdly, despite high overall urbanization rates, there is a trend towards smaller household sizes and declining urban population densities due to sprawling suburbs. This is very likely associated with more individualization of travel demand, i.e., mobility patterns are likely to be spread out more in space and time in the future. This favors low-capacity vehicles over high-capacity vehicles and may thus go hand-in-hand with vehicle automation. This is because replacing drivers makes more economic sense for low-capacity vehicles where driver costs represent a larger share of the total costs relative to large vehicles. Overall, against the background of these demographic trends, Japan and Germany appear to be predestined to move forward on the path of vehicle automation, as their aging societies are likely to benefit from its potential in many respects.

4 Road Transport—Policy, Governance and Regulation

4.1 Germany

Responsibilities for planning, building and maintaining the public road network in Germany closely follow the political structure of the Federal Republic. According to the Federal Trunk Road Act (FStrG)—officially enacted in 1953—federal trunk roads are public roads that form a coherent transport network and are intended to serve long-distance transport [15]. A distinction is made between federal motorways (usually grade-separated) and federal roads. Infrastructure expansion planning for federal trunk roads is partly embedded in European infrastructure development planning. The core strategic document is the Federal Transport Infrastructure Plan (Bundesverkehrswegeplan, BVWP), which covers all investments by the Federal

Government in its transport infrastructure, not only construction and expansion, but also maintenance and renewal [21]. Since 2005, federal highways are subject to road tolls for heavy goods vehicles. Since then, the toll system has been gradually expanded to cover substantial parts of the remaining federal roads network. Beyond this, the initially introduced restriction of road tolls to motor vehicles or vehicle combinations with a total permissible weight of 12 tons or more has been lowered to 7.5 tons since October 2015. A toll for passenger cars in Germany was discussed since the 1990s and formally introduced on January 1, 2016. But the respective law did not come into force because the European Court of Justice (ECJ) held that it is incompatible with European Union (EU) law.

Responsibilities for the remaining road network are regulated by state road laws. As a general rule, the responsibility for road construction and maintenance for state roads lies with the respective federal state, for county roads with the respective county, and for local roads with the respective municipality. Private road infrastructures and public private partnerships (PPP) in road infrastructure construction and operation are a matter of ongoing political debate. So far, only a very small number of projects have been realized.

The legal framework for traffic regulations, liability, penalties and fines, driving suitability, vehicle register and driver's license register is provided by a federal law, the Road Traffic Law (StVG). This was first introduced in 1909, and more than 100 years later the current law still contains much of the original version, exemplifying the longevity of regulatory frameworks despite fundamental technology transitions in road transport [18]. The most important specific provisions for the use of public roads as well as the 'rules of the road' are contained in the Road Traffic Regulations (StVO) [17]. The StVG thus delegates the creation of concrete regulations to ensure the safety and ease of road traffic to the level of legal ordinances. In addition to the StVO, the Federal Government has issued a General Administrative Regulation on the German Road Traffic Regulations (VwV-StVO) [23]. These contain explanations for the competent authorities on the practical interpretation of the StVO. The VwV-StVO do not themselves have the quality of a legal norm, but they do bind the authorities. This is particularly important when it comes to the application of discretionary regulations, which is the case with a number of the central regulations of the StVO. This construction—and specifically the overall stated objective of Road Traffic Regulations to ensure safety and ease of road traffic—has triggered a broader debate on whether it favors motorized road vehicle use compared to other modes of transport, and creates legal obstacles to a transportation policy aiming at more sustainable and safer mobility options [93]. StVG and StVO are regularly amended. Two amendments of the StVG in 2017 and 2021 were specifically dedicated to address topics that emerged in the context of connected and automated driving (CAD)—to enable the integration of automated driving as a new principle into the regulatory framework for road traffic (2017), and to support the deployment of fully automated vehicles on German roads (2021).

4.2 Japan

In Japan, the Road Law stipulates the road significance and type, management entities, procedures of route designation, certification and abolition, and allocation of cost burden required for road management. The Road Law specifies four types of roads for public traffic: National Expressways, National Highways, Prefectural Roads, and Municipal Roads. Generally, the Minister of Land, Infrastructure, Transport and Tourism (MLIT) is the road administrator for National Expressways and National Highways within designated sections. For National Highways outside the designated sections and Prefectural Roads, the prefecture or designated city is the road administrator. The respective municipalities are the road administrators of Municipal Roads. An expressway company may, with the permission of the MLIT, construct or improve toll roads and act on behalf of the road administrator for certain authority. In 1987, an arterial high-standard highway network plan was approved to develop National Expressways and National Highways with access control of approximately 14,000 km, and the progress rate at the end of fiscal year 2021 reached approximately 87%. The approximately 1,100 km expressway between Tokyo and Fukuoka constitutes the Asian Highway Route (AH1) that was adopted by the United Nations Economic and Social Commission for Asia and the Pacific (ESCAP) [90].

In the 2013 revision of the Road Law, a road inspection cycle was established from the viewpoint of preventive maintenance, based on the current situation of aging road infrastructure that was constructed predominantly during the post-war high economic growth period of the 1960 and 1970s. The enforcement and guidance for appropriate routing of heavy vehicles that cause road deterioration are being strengthened.

The Road Traffic Act stipulates measures to prevent danger on roads, and ensure the safety and smoothness of other traffic. Specifically, it stipulates the driver's license system, traffic rules, penalties, and fines. The Road Transport Vehicle Law stipulates automobile ownership, safety regulation, and a periodical vehicle inspection system. The Road Transportation Law and Freight Car Transportation Business Law regulate the automobile transportation business.

In recent years, the Road Traffic Act, under the jurisdiction of the National Police Agency, and the Road Transport Vehicle Act, under the jurisdiction of Road Transport Bureau of MLIT, have been amended to enable automated vehicles to travel on public roads (for a more detailed discussion concerning this legal reform, see chapter "Governance, Policy and Regulation in the Field of Automated Driving: A focus on Japan and Germany"). In addition, the embedded facilities that support automated driving have been added to road accessories in the revised Road Law.

4.3 Comparative Conclusions

Despite many differences in detail, the regulatory frameworks of the road infrastructure in both countries share many similarities. There is a hierarchical road network

Table 2 Key figures on the physical extent of the federal highway infrastructure [19, 89]

	Germany	Japan
Federal/National highway network length (km)	51,027 (2017)	52,243* (2020)
Federal/National highway network km per capita	0.0006 (2017)	0.0004* (2020)
Federal/National highway network km per km^2	0.143 (2017)	0.138* (2020)

* Roads less than 5.5 m in width have been excluded from the statistics

system ranging from federal highways to municipal roads, with responsibilities for the respective administrative levels. At least for the federal road network, there is high-level and long-term planning looking several decades ahead. In combination with the enormous financial investment and sunk costs of building the infrastructure, this is one important component of the inertia of the transport system when it comes to all aspects associated with the infrastructure hardware. This emphasizes potential barriers to vehicle automation if alterations or upgrades of the physical infrastructure are required in order to enable vehicle automation. As for the organization of road traffic through road traffic regulations, it should be noted that some parts of the regulations are even older than large sections of today's road network. Hence, there is also substantial inertia in the regulations. Nevertheless, both countries have taken the first regulatory steps to enable automated driving, illustrating the ambition to at least overcome the regulatory barriers to automation as soon as possible.

5 Road Transport—Infrastructure Supply

5.1 Comparative Statistics

See (Table 2).

5.2 Germany

The modern German road network—urban and interurban—was predominantly shaped in the 20th century. While in some cases modern roads were built on or extended street layouts from ancient or medieval periods, the majority of today's roads are a result of 20th century planning. This applies to urban roads as a result of the spatial extension of settlement areas in the last century, as well as to interurban roads, as the planning of the modern interurban road network was ignited by the advent of the automobile. The layout and design of the vast majority of streets in Germany today reflects post-WWII planning [106].

In transport planning in Germany from the 1950s through to the 1970s, the car was clearly center stage, and the paradigm of the "car-oriented" city dominated urban infrastructure planning. In the central cities, parking on historic market places or other public spaces was commonplace. There were widespread plans for high-capacity urban roads, in many cases freeways and even freeway intersections in central neighborhoods of major cities such as Munich or Berlin [120]. Likewise, plans for suburban neighborhoods aimed to facilitate vehicular travel. Specifically, during the 1960s, when these paradigms dominated and funding for road construction was abundant, and during the 1970s, when the earlier plans were carried out, such plans were implemented. Much of the heritage of the era of auto-oriented planning such as inner-city parking structures and urban freeways, which still largely shape automobile infrastructure in German cities today, originate from that era. Examples that epitomize both the philosophy of auto-oriented planning in the 1960s and the paradigm shift that has taken place since then, are elevated freeways in the city of Ludwigshafen (built in the 1970s and torn down from 2020 onward [2]), and a city-center parking garage in Aachen (built in the 1960s and torn down in 2021 [111]). Since the 1980s, planning paradigms gradually changed. Examples are pedestrian areas in city centers which became commonplace in the 1980s, and a reinforced focus on cycling infrastructure from the 1990s. Since 2007, official German urban street design guidelines place a higher value on the quality of the urban space than on vehicular flow [42].

Today, German cities are characterized by a street infrastructure that includes a mix of design elements ranging from the era of auto-oriented planning to today's planning paradigms. Despite German expert self-assessment of road infrastructure which ranks Germany only on position 20 globally with regard to road quality [119], it is fair to say that the automobile infrastructure is relatively good in Germany. Drivers find favorable conditions, which are exemplified by many interurban roads which provide for high driving speeds, a lack of a general speed limit on the "Autobahn", and an ample supply of parking except in dense urban neighborhoods. Parking policies (as well as parking costs) also represent a good example for the continued policy paradigm of accommodating cars in cities rather than banning them: While public parking spaces are gradually removed in numerous urban areas, parking requirements provide for a growing inventory of private parking at the same time [112]. Travel mode-specific accessibilities are another indicator for good conditions for driving relative to using other modes: With regard to the number of inhabitants one can reach within a given travel time budget from almost any location in Germany, including central cities, the car outperforms other modes of travel by far. In simpler terms: The car is mostly the fastest and most convenient mode of passenger travel in Germany.

5.3 Japan

The intercity road network and major urban roads in Japan were formed after WW II. For topographical reasons, some intercity roads follow the same routes as ancient trunk roads. From 1940, inspections of the existing roads were carried out as well as assessments of the needs for road infrastructure. These formed the basis for road development planning, and most of the road infrastructure currently in use was constructed after WW II.

In 1957, the number of registered automobiles exceeded 2 million, although this number was only approximately 130,000 just after WW II. Thus, the era of motorization had arrived in Japan. However, the road infrastructure at that time was so poor that only 23% of the first-class national roads were paved [89]. Ralph J. Watkins, who was invited by the Japanese government to investigate and plan the Meishin Expressway in 1956, stated in his report, "The roads of Japan are incredibly bad. No other industrial nation has so completely neglected its highway system."

In 1953, the "Act on the State's Tentative Financial Measures for Road Construction Projects" was enforced, and fuel and other automobile-related taxes were consequently used for road development. As a result, stable financial resources were secured and five-year road development plans, from the 1st to the 11th, were created and executed. Thus, significant nationwide road development was achieved for more than 50 years. During this period, major roads have been paved and over 10,000 km of expressways have been developed nationwide.

As of April 2021, 12,082 km of the Arterial High-standard Highway, which is 86% of the planned approximately 14,000 km extension, has been developed in Japan. However, when compared internationally, Japan's road infrastructure is inferior [57], especially high-standard highways. First, the network is discontinuous, and many sections have been provisionally constructed with two lanes. In other countries, expressways have 4 and 6 or more lanes for approximately 75% and 25% of their length, respectively. However, in Japan, 38% of the length is 2 lanes, and the extension of 6 lanes or more accounts for only 6%. Second, the travel speed is low. A comparison between the average speed on roads connecting major cities reveals speeds of 80 km/h or more in European countries, 95 km/h in Germany, and 60 km/h in Japan. This is partly due to speed limits; the general speed limit on interurban roads in Japan is 100 km/h; however, this varies between 80 and 120 km/h, according to local conditions. An additional reason for low travel speeds in Japan is presumably the discontinuous sections of expressways, provisional two-lane expressways, and sections with steep gradients and topographical restrictions that have low speed limits. Third, there are many tunnels and bridges, resulting in high construction and maintenance costs. Approximately 12.9% and 15.8% of the length of expressways in Japan consist of tunnels and bridges, respectively, which are expensive to construct and maintain. Approximately 50 years have passed since the construction of this infrastructure, and large-scale renewal is expected to be required in the near future.

Conversely, from a global perspective, the advantages of Japan's expressway infrastructure are as follows. First, a nationwide unified Electronic Toll Collection (ETC) system has been introduced and expanded, and one on-board unit can be used even when different expressway company routes are used. In addition, travel demand management (TDM) through toll rate measures has become easier. Second, rest facilities provide a high standard of service. Since the privatization of the Japan Highway Public Corporation in 2005, expressway companies have endeavored to improve the service level of rest facilities. In addition, as a disaster prevention base, the number of expressway rest facilities that also support activities such as rescue and medical care in the event of a large-scale wide-area disaster is increasing.

5.4 Comparative Conclusions

As for the quantity and quality of the road network there are noteworthy differences between the two countries. This particularly applies to the federal interurban highway network, where use of vehicle automation is likely to become widespread first. Relative to the population, the German federal highway network is denser, allows for higher driving speeds and there is no tolling system implemented so far. These are factors which may contribute to a stronger motivation to move towards automated vehicular flow in Japan: The infrastructure, which—due to the country's topography—is characterized by tunnels and bridges and is therefore much more expensive to construct and maintain, is more constrained when it comes to capacity and cannot easily be extended. Vehicle automation—if implemented as connected and cooperative automation—comes with the promise of making better use of existing infrastructure capacity. Due to the restrictions described above, the Japanese expressway infrastructure may thus benefit more from automation than Germany's. Moreover, with lower travel speeds due to speed limits and an electronic tolling system in place, there is already stronger traffic control in place in Japan relative to Germany. However, the small number of lanes and narrow shoulders on expressways may hinder the spread of automated driving in Japan. Hence, Germany may not only benefit less with regard to traffic flow from automation, it may also have a longer journey to transition from today's human driving habits to automated homogeneous vehicular flow.

6 Public Transport and Transport Services—Policy, Governance and Regulation

6.1 Germany

Planning, operation and financing of public passenger transport in Germany are organized in a comparatively complex manner. For a better understanding, it is necessary to differentiate with regard to the mode of transport and the responsible governance level within the context of a federal republic integrated into the structures of the EU.

Long-distance rail transport is dominated by Deutsche Bahn AG (DBAG), a company fully-owned by the Federal Republic of Germany and organized under private law. It is provided on an economically self-sufficient basis [41]. Other domestic and foreign railroads are in principle free to offer long-distance passenger services. However, this has so far been practiced only to a modest extent. Potential competitors regularly complain about high barriers to market entry. As an infrastructure company, DBAG also operates around 87% of the German rail network [19]. The federal states are responsible for organizing regional rail transport. In their local transport laws, they have assigned this function either to themselves or to municipal special-purpose associations. These in turn contract and finance the local transport services in cooperation with the respective state planning authorities [24]. A whole range of companies, which may be privately- or publicly-owned, act as providers of these ordered services.

The legal basis for local road passenger transport in Germany is the Passenger Transport Act (PBefG), which applies to the paid or commercial transport of passengers by streetcar, trolleybus and motor vehicle [16]. According to this law, for-profit services and specifically scheduled public transport supply are subject to approval. The responsible authorities are the counties or cities. As a rule, these services are operated by county- or city-owned public transport companies, sometimes also by private transport companies. In practice, they use a wide range of means of transport, such as streetcars and local, city and regional buses, and in large cities often also subways or light rail systems. In addition, special forms such as dial-a-ride buses or dial-a-ride shared cabs, and "exotics" such as suspension railways or people movers are also included in public transport. The PBefG was amended in 2021 in order to accommodate new forms of mobility services like private- or shared vehicles for hire if they replace, supplement or consolidate conventional public transport services [129].

Most local public transport authorities are also organized in transit districts (Verkehrsverbünde) or tariff associations, in which all means of transport can be used with one ticket. These are special-purpose associations to which the public transport authorities have delegated certain management tasks, for example, the preparation of local transport plans, in addition to the coordination of fares [65]. The legal basis is provided by the relevant local transport laws of the federal states. Some of them have made these districts or associations mandatory by law, while others leave their formation to voluntary decisions of public transport authorities [103]. Services and payments in passenger transport are often regulated in a transport contract.

The distribution of responsibilities among different actors in a multilevel governance system [104] such as that which has emerged in Germany does not always lead to satisfying results, especially when origin–destination patterns cross jurisdictional boundaries. Cost- and revenue allocation, different interests, and different political power structures (including different transport policy priorities) create tensions and coordination challenges between organizations working at the same level of governance within different jurisdictions (horizontal coordination). The same holds true for vertical coordination, since scale differences may drive different perspectives, objectives and priorities.

6.2   Japan

In order to understand the characteristics of public transport in Japan, the underlying paradigm of private management and fare revenue as the economic basis for the services is important. In Japan, there is a strong belief that transport is a private business that can make a profit and loss, apart from the state-owned railways and the state-owned airlines, which were privatized in the 1980s.

Japan National Railway (JNR) was established as a public corporation in 1949, but was unable to keep up with changes in the industrial structure associated with rapid economic growth from the late 1950s, and fell into a deficit from 1964; in 1986, a bill related to JNR reform was submitted to parliament, and in 1987, division and privatization were implemented [83]. Excluding the impact of COVID-19, the listed Japan Railway (JR) companies are profitable. In addition, the private railway companies have generated revenues through diversified management, including real estate and lifestyle service businesses.

Against this background, the idea that it is better for transport to make a profit as a private-sector business is strongly entrenched in Japan. This is known as the "independent profit-making system" and is the basic concept of public transport in Japan. Laws were amended in the early 2000s to bring more market principles into transport. From the perspective of enabling operators to develop their business flexibly based on management decisions and to improve the efficiency and vitality of their business, the Railway Business Law was amended to ease the freight railway business exit from a permit system to a prior notification system (2000), and the Road Transport Law was amended to ease bus and taxi entry from a license system to a permit system, and exit from a permit system to a prior notification system (charter buses in 2000, passenger buses and taxis in 2002) [91].

Another characteristic of public transport in Japan is its vertically divided organization, which impacts on the coordination between municipalities. In recent years, local authorities have increasingly become the operating body for local transport. However, there are many problems, such as routes that do not cross the boundaries of municipalities. With privatization and an increased focus of the management on profitable services, there has also been a decline in the number of buses in rural areas, and a withdrawal of bus services. Increasingly, local authorities are taking the lead in operating community bus services after private operators have withdrawn [92].

The background of more municipalities beginning to introduce community buses is the Law on the Revitalization and Regeneration of Regional Public Transport (2007), which sets out a system to promote the formulation of a "comprehensive regional public transport coordination plan" to be initiated by the municipalities. However, many local authorities have difficulties responding to the given responsibilities in local transportation planning, in terms of inadequate institutional structure, personnel, and budget [117]. Besides, regional associations such as the Transport Union (Verkehrsverbund) in Germany have not flourished in Japan. As a result, many community buses are set up with routes that do not straddle the boundaries of the municipalities, which does not always match the user flow-lines. It remains to be seen whether this law will function effectively.

6.3 Comparative Conclusions

Regarding the organization and regulation of public transport, the main difference between Germany and Japan refers to the role of private companies and the belief in market forces when it comes to the provision of public transport. In Germany—as in much of continental Europe—there is a historically-rooted and deeply ingrained belief in state intervention in transport, and specifically public transport. Despite deregulation and a simulation of competition through a complex system of tendering and procurement, public transport is essentially a state-run and heavily subsidized enterprise. One could say that in Germany the starting solution for public transport provision is that of a public enterprise with some market elements. In Japan, it is the opposite: public transport—as other domains of the transport sector—is a private business with some state intervention in those cases where it does not work without. As for vehicle automation, these different approaches in public transport provision raise different expectations regarding the introduction of automated public transport services: It seems likely that private stakeholders will be more dominant in initiating automated public transport services in Japan than in Germany, where the activities of the public sector are likely to play a bigger role.

7 Public Transport and Transport Services—Infrastructure and Supply

7.1 Comparative Statistics

See (Table 3).

Table 3 Key figures on the physical extent of the rail network in Germany and Japan [19, 29, 45, 113]

	Germany	Japan
Total rail network length (km)	38,394 (2019)	27,311 (2015)
Total rail network per capita (km/capita)	0.00046 (2019)	0.00021 (2015)
Total rail network km per km^2	0.11 (2019)	0.072 (2015)

7.2   Germany

Modern public transport in Germany originated in the 19th century when rail transport—urban and interurban—started to be widespread. Hence, in its early days, public transport was mostly rail-based and initiated by private investors. From the beginning of the 20th century, the state took over, and since the mid-20th century many tramway systems—which at their peak were also present in smaller cities—stopped operating and were replaced by buses or high-capacity urban rail such as subways and metro rail [47]. The latter, however, never reached the territorial coverage that tramways once offered, and are concentrated in high-density urban areas and corridors.

Today, rail-based public transport—commuter rail, subways and tramways—represents the backbone of public transport in large cities. All except four German cities over 200.000 population have rail public transport [46]. Smaller cities, rural areas or the low-density areas of large cities are usually catered for by buses. Since the 1990s there is also a tendency to re-introduce (e.g., Saarbrücken) urban tramways, or re-extend remaining tramway networks (e.g., Karlsruhe) in German cities. This, however, takes place at a much lower scale compared to other European countries such as France or Spain. There are also many examples where the re-introduction of urban tramways—often favored by planners and politicians—were declined in local referendums [46].

In large German cities, specifically those areas with rail public transport, the quantity and quality of public transport in terms of travel times and service frequency can generally be regarded as good. This is exemplified by the public transport modal split, which in large cities (population over 100.000) with urban rail is about twice as high as in cities without. Despite good coverage of public transport, even in large cities—and of course even more so in rural areas—travel speeds by public transport are in most cases well below those of the car [46].

As in other parts of the world, new flexible transport services based on street motor vehicles—mostly shuttle buses—have emerged in Germany. There are two strands to this development: First, conventional—often state-owned—public transport providers sought new options to provide public transport in low demand areas. In such areas, ridership has declined in recent years due to increased car ownership among seniors, and fewer school students due to demographic change. The cost of offering basic public transport services in such low-demand areas has increased. Flexible transport services which operate on request and often utilize smaller vehicles are believed to be a solution in such a situation [58].

Secondly, private entrepreneurs hoped for a business case by offering flexible transport services in high-demand urban areas in Germany. This was inspired by business models of Uber or Lyft in the US, where private individuals offer for-profit services. These business models are not allowed in Germany due to public transport regulation (even though Uber exists as a brand in Germany, it offers a different kind of service). Nevertheless, in recent years new flexible, on-demand shuttle-like services emerged in various German cities (e.g., Berlin, Hamburg, Hannover) as an addition to existing high-capacity urban transport [27, 94]. With regard to user cost, these services usually sit between conventional public transport and taxis. While there is now—after amendments to the public transport regulation—a long-term legal perspective for such services, it is unclear if they are economically sustainable and meet expectations with regard to profitability [59].

7.3 *Japan*

As for public transport in Japan, before WW II, passenger transport consisted mainly of railways and trams. With post-war reconstruction, trams and buses became the main modes of transport. With motorization beginning at the end of the 1950s, and the 1964 Tokyo Olympics, trams were closed to make space for cars. In large cities with populations of one million or more, many of the tram lines became subways. Around 2000, trams began to be re-evaluated, and in some cases, such as Toyama City, the old, disused freight railway line was revived as tram, but it has been difficult to revive trams once they have been closed due to opposition from car users and other factors.

The number of public transport passengers is divided between the three metropolitan areas of Tokyo, Osaka and Nagoya, and other regional cities: in 2010, 70% of commuters in the Tokyo metropolitan area used public transport (railways and buses), and 10% used private cars. In rural Tottori Prefecture, however, public transport accounted for 7%, and private cars 74%. The national average is 33% for public transport, and 48% for private cars [121].

Deregulation around 2000 caused transport operators to withdraw from unprofitable routes, forcing local authorities to invest taxpayers' money to maintain mobility for daily life. However, many local governments lack the skills to manage public transport, making it a difficult task to maintain travel services for vulnerable groups who cannot drive. In towns, villages and rural areas where fixed-schedule buses are not economical, on-demand buses have been introduced in some cases, operated by local authorities according to user-demand. However, as public transport in the rural region is not a profitable business to begin with, the more passengers the on-demand bus service attracts, the more it puts pressure on the local government's finances.

Private transport services for profit such as Uber have not yet been authorized due to concerns about competition with taxi services in Japan. In addition to lobbying by the taxi industry to MLIT and the Parliamentary Association for Taxis, this is due to the idea that public transport in Japan is regarded as a private profit-making business, but that its safety should be guaranteed by the Japanese government. The idea that the government guarantees the safety of passengers is different from, for example, the US idea that safety is the responsibility of the passenger because they choose to use convenient services.

One issue specific to Japan is the shortage of bus drivers. Bus driving has become an occupation avoided by young people because of the lower-than-average wages, irregular working hours and the need for a Class 2 driver's license. There is concern that the shortage of bus drivers will accelerate in the future due to the aging of the existing bus driver population. This major social problem could be alleviated if automated driving technology improves, and the regulations are relaxed so that a regular driver's license holder can become a bus driver.

As mentioned above, public transport in Japan is based on an independent profit-seeking system. Although this may lead to improved service levels through competition, the reality is that increasing car ownership makes it difficult to continue profitable business, especially in rural areas. The voluntary curfew due to COVID-19 and the generalization of telework and online conferencing have dealt a heavy blow to public transport services, raising the question of whether and how to maintain and expand public transport, as well as the pros and cons of an independent profit-making system.

7.4 Comparative Conclusions

Despite differences in the underlying philosophies—i.e., public and state-controlled vs. private for-profit provision of public transport services—there are many analogies between the status quo, trends and challenges as regards public transport supply in Japan and Germany. Rail-based services are the backbone of public transport in major metropolitan areas, but in many smaller cities rail-based services—specifically tramways—were dismantled in the mid-20th century. Despite a recent tendency to re-introduce and extend urban rail transport to address urban transport problems, there are substantial barriers to re-introduction. However, in rural areas, providing public transport represents a major challenge, and the emerging lack of bus drivers is an additional problem. Against this background, there are two important conclusions as regards vehicle automation: Firstly, public transport providers in rural areas would benefit substantially from automation as it may help to lower the cost of providing public transport while at the same time solving the staffing challenge. Secondly, the perspectives regarding the balance of rail-based and road-based public transport in metro-areas are unclear. The question is whether road-based public transport with automated vehicles will offer a suitable alternative to rail-based transport as regards comfort and capacity, without exacerbating environmental problems. If so, this may curb the recent trend to urban rail and reinforce the focus on road-based urban public transport, possibly even with smaller and more flexible vehicles.

8 Pricing, Taxation and the Cost of Transport

8.1 Comparative Statistics

See (Table 4).

Table 4 Key figures on the cost of and expenditures for transport by car and public transport, Germany and Japan [14, 19, 28, 33, 39, 40, 79, 84, 97, 123, 124]

	Germany	Japan
Average annual vehicle tax per passenger car (local currency unit, LCU)	158€* (2016)	39,500 JPY** (2019)
Average annual vehicle tax per passenger car as percent of GDP per person per day (%)	152%* (2016)	325% ** (2019)
Cost per liter of gasoline (LCU/liter)	1.42€ (2019)	147.40 JPY (2019)
Cost per liter of gasoline as percent of GDP per person per day (%)	1.24% (2019)	1.21% (2019)
Tax as proportion of total gasoline price (%)	60% (2020)	36.5% (2019)
Average user cost per passenger km of public transport (excl. Taxis) (LCU)	0.11€ (2017)	16.6 JPY*** (2018)
Average user cost per passenger km of urban public transport as percent of GDP per person per day (LCU)	0.1% (2017)	0.1% (2018)
Total expenditure for transport and communication as proportion private household expenditure (%)	16% (2019)	17% (2019)

* The annual vehicle tax for passenger cars in Germany depends on displacement and CO_2-emissions; the average annual vehicle tax as presented here is the average expenditure per vehicle based on an income and expenditure survey
** The annual vehicle tax for passenger cars in Japan depends on displacement and environmental performance; the average annual vehicle tax as presented here is the amount for the vehicle whose displacement is over 1,000 cc and not more than 1,500 cc. The amount is reduced depending on the environmental performance. The value does not include the amount of tonnage tax
*** Calculated using the values of passenger-km and operating income for private charter buses, scheduled buses, JR, and private railways
**** *Calculated using real income and expenditures for transport and automobile related cost in worker households

8.2 Germany

During recent decades, Germans have been spending a relatively stable proportion of about 15% of their disposable incomes on transport [33] with similar proportions of income spent on transport by households of different income classes [31]. In 2013, the average transport budget amounted to 350 Euro per household per month, 90% of which was spent on automobile travel. This illustrates that ownership and use of cars constitute the most important factor when it comes to the total cost of mobility.

In 2016, the average monthly total cost of ownership (TCO) of a passenger car in Germany (including used and new vehicles) was 310 Euro. Depreciation, other fixed costs (vehicle tax, insurance, repair and maintenance) and fuel each constituted about a third of this total cost [39]. In Germany, the costs of fuel and parking represent the only marginal cost of automobile use (road pricing is a rare exception). There is little knowledge about parking expenditure in Germany, but it is evident that relative to neighboring high-income countries, parking is relatively inexpensive in German cities. In rural and suburban areas drivers usually face no costs for on-street parking. Overall, parking does not add much to the average cost of driving in Germany, and the marginal cost of driving is mostly defined by fuel costs [66].

In 2016, given a monthly mileage of about 1100 km, 310 Euro per passenger car per month resulted in total costs per passenger car vehicle km of about 30 Euro Cents per km, or marginal (fuel) costs of about 10 Euro Cents per km [39]. Considering a passenger car occupancy rate of about 1.4 this leads to car usage costs per passenger km of about 21 Euro Cents TCO, or about 7 Euro Cents with regard to marginal costs. These marginal costs are below the average cost for public transport use in Germany of about 10 Euro Cents per km (after factoring in all discounts, season tickets and single fare tickets). The costs for all other motorized modes of transport (taxis, emerging mobility services, rental scooters etc.) are substantially above that price level (50 Euro Cents per km and more) [66].

Depending on current fuel costs, taxes constituted about 50–60% of the fuel price in the last decade [1]. The largest tax component of fuel price (energy tax plus CO_2-emission levies) is a fixed amount per liter (Gasoline: 73 Euro Cents/liter; Diesel: 55 Euro Cents/liter) and only the absolute amount of the value added tax (19%) fluctuates with the total fuel price. This also means that—after accounting for inflation and increased vehicle efficiency—real tax income from fuel taxes actually declines over time. This problem is aggravated by the emergence of electric vehicles which currently pay substantially less use-related taxes and levies (only electricity tax). Hence, the introduction of road pricing in Germany in the next few years appears a likely option to compensate for the fuel tax income foregone.

While fuel taxes constitute an important source of state revenue (and are about twice as high as the annual state budget spent on the road network in Germany) [19], public transport is heavily subsidized: About 30% of the operational costs and about 50% of the total costs of public transport are covered by subsidies [11]. Hence, the total cost of public transport of about 10 Euro Cents per km is a result of strong subsidization.

According to the German consumer price index development over the last 30 years, overall transport costs increased somewhat more strongly than the average consumer price index [34]. This is mainly due to the fact that the cost of public transport usage and fuel increased above average consumer price trends. In contrast, the growth of the costs of passenger car purchases was below average price increases. Hence, overall public transport consumer prices increased more than passenger car use, with a shift in the cost of passenger car use from fixed costs to marginal costs.

8.3 Japan

In Japan, automobile taxes are imposed at multiple stages in the lifetime of vehicles: acquisition, ownership, and actual travel. At vehicle acquisition, a consumption tax and an automobile tax or a light automobile tax (environmental performance discount) are levied; during ownership, class-based automobile taxes/light automobile taxes for mini cars, and tonnage taxes apply; and, when vehicles are driven, there are multiple fuel taxes: gasoline tax and local gasoline tax for gasoline, diesel oil delivery tax for diesel, and oil and gas tax for liquefied petroleum gas. Of these, the environmental performance discount was introduced in 2019 to replace the older automobile acquisition tax; the old tax was assessed on acquisition price at the time of vehicle purchase, but the new tax, which applies to used as well as to new vehicles, is based on acquisition price but also takes fuel efficiency and other factors into account. The tonnage tax is based on gross vehicle weight at time of vehicle purchase and subsequent inspections. The class-based automobile taxes/light automobile taxes are assessed yearly based on vehicle emissions. For many years, the tonnage tax, together with fuel taxes and the now-abolished acquisition tax, were earmarked for use by the national and municipal governments primarily to maintain and construct general-purpose (non-highway) roads, but this system was discarded in 2009, and today the proceeds of these taxes are treated as general-purpose funds. Usage fees for highways are collected separately from these taxes. In general, roads in Japan are open to the public and free of charge; however, for a brief period in the mid-20th century, stretching from Japan's post-war recovery through its high-growth period, an exception was made in the form of a toll-road system to raise funds for building and maintaining a trunk network of roads connecting major cities at high speeds (see Ministry of Land Infrastructure Transport and Tourism (MLIT) [85]).

According to the Japan Automobile Manufacturers Association, initial budgeting for fiscal year 2021 shows that total tax revenue collected from automobile users in Japan—including consumption taxes on vehicle acquisitions and fuel purchases— amounted to 8.6 trillion JPY, or 8.7% of the nation's overall tax revenue of 99.3 trillion JPY [54]. As an example, to illustrate the scale of these taxes, the acquisition of one standard-sized car and its 13-year ownership would entail an automobile tax (environmental performance discount) of 21,700 JPY, a consumption tax of 410,400 JPY, class-based automobile taxes of 504,000 JPY, and a tonnage tax of 196,800 JPY [55].[1]

Let us compare the costs borne by Japanese automobile users against the cost of public transportation: According to the Annual Report on the Family Income and Expenditure Survey from the Statistics Bureau of Japan, the average monthly automobile-related cost (including purchase cost) in a household was 19 thousand JPY in 2019. Average monthly distance traveled per privately owned passenger car is calculated to be 729 km in the fiscal year 2019 [87]. Considering the passenger car occupancy rate of 1.41 in 2015, and the number of passenger cars owned by

---

[1] Conditions for calculation are as follows: Displacement: 2 L; vehicle weight: less than 1.5 t; fuel economy under JC08: 20.4 km/l (CO_2 emission: 114 g/km); vehicle body price: 2.42million JPY.

a family of 1.04 (Automobile Inspection & Registration Information Association [6] at the end of FY2019, the average cost per person-kilometer of privately-owned passenger cars is roughly estimated to be some 17.4 JPY. In comparison, the average user-cost per passenger km of public transport, including buses and railways, costs 16.6 JPY, and thus from a cash-burden perspective vehicle travel costs slightly more than public transport travel. However, vehicle travel cost is likely to be even higher. It has been pointed out that the cost of purchasing a car is underestimated in the Family Income and Expenditure Survey. This is one of the reasons. Note that public-transit facilities in Japan, even if operated by public institutions, are intended in principle to be independent and self-supporting based on fare revenue; however, the reality is that many public-transit networks in rural areas of Japan are unsustainable without national or local government support. Indeed, according to 2019 data, 69% of Japanese transit-bus operators are unprofitable, and some 3.5% of all transit-bus routes (measured by distance) have been eliminated since 2007 [86].

8.4 Comparative Conclusions

Despite many differences between Germany and Japan as regards pricing private vehicular travel and setting prices for public transport (e.g., through subsidization) there are relevant analogies: Firstly, depreciation—or the cost vehicle purchase respectively—represents only one component of the costs of private motoring, which comprise a wide range of different fixed and variable cost components. Hence, an increase in vehicle purchase costs due to automation technology is not linearly an increase in the cost of driving. Many other factors come into play and—depending on other framework conditions—operating automated vehicles may not be much more costly than conventional vehicles. Secondly, the car, i.e., the mode that today often offers the most individual comfort and higher overall travel speeds in most everyday travel situations except in high density urban environments, costs as much or more on a per-km basis than public transport. The population of travelers is very heterogeneous and some are capable of opting for high-cost high-speed modes, while others rely on low-cost modes with lower travel speed and comfort. Today's mode use patterns emerge as a result of the interplay between these properties of the transport system and the population's needs and economic capabilities. It remains to be seen how automated vehicles will be positioned in this spectrum of cost, speed and comfort in a future transport system, and which socio-economic groups in the population will benefit the most from vehicle automation.

Table 5 Key figures on the economic relevance of the machinery and transport equipment industry [26, 128]

	Germany	Japan
Machinery and transport equipment (value added, % of GDP)	9% (2018)	9% (2018)
Number of passenger cars produced per 1.000 population 1961	25 (1961)	3 (1961)
Number of passenger cars produced per 1.000 population 1991	58 (1991)	79 (1991)
Number of passenger cars produced per 1.000 population 2019	56 (2019)	66 (2019)

9 Automotive Industry as an Economic Factor

9.1 Comparative Statistics

See (Table 5).

9.2 Germany

In Germany, the contemporary automotive industry is a key industrial sector and a major economic factor. The companies Daimler and Benz, both located in today's southern German state of Baden-Württemberg and combined in the 1920s to Daimler-Benz, were among the first companies to start industrial production of automobiles at the end of the 19th century. Volkswagen, today the second largest automobile manufacturer worldwide after Toyota, was founded in 1937 by the German Nazi regime. Volkswagen production started shortly thereafter in the rural municipality of Fallersleben, which today is part of Wolfsburg, a city that developed around the Volkswagen production plant in what today is the northern German state of Lower Saxony. Today, the two regions, southern Germany and parts of Lower Saxony, are still epicenters of the German automotive industry. However, the automotive industry—including automobile manufacturers and suppliers—is a key player for the regional economy almost everywhere in Germany. For example, in 2016 there were a total of 41 final assembly plants distributed throughout Germany [12].

Before the COVID-19 pandemic, in 2019, the total turnover of the German automotive industry was about 440 billion Euros, equivalent to about 13% of the German GDP or almost 25% of the total industry turnover in Germany [35]. Vehicle manufacturers account for about three quarters of this automotive industry turnover. Consequently, the automotive industry is of paramount importance for the German job market. In 2016 the automotive industry employed about 800.000 individuals directly; about 2% of the German workforce [12]. About the same number of employees depend indirectly on the automotive industry, e.g., in the chemical or textile industries. With this relevance of the automotive industry for the job market,

Germany clearly stands out in Europe: almost half of the European jobs that directly or indirectly depend on the automotive sector are to be found in Germany.

Germany's automotive industry is strongly and increasingly international in both production and sales, and its growth is driven by the global market: While in 2008 about half of the 11 million cars produced by German manufacturers were produced in Germany, this figure reduced to about a third in 2016, when German manufacturers produced about 16 million vehicles. Likewise, the proportion of German vehicles sold abroad increased from 80 to 85% in the same period. Thereby, German car makers—specifically the brands Mercedes-Benz, Audi and BMW—specialize in high-margin luxury vehicles. As for sales in this vehicle segment, German manufacturers have a large lead in the global car market, and much of the recent increase of the automotive industry turnover was driven by the growth of SUV sales [12].

As a consequence of this strong position of the German automotive industry and its key role for the national economy, it is no surprise that automotive industry lobbying is a major influential factor in German transport and industrial politics [9]. For example, there is a long history of lobby intervention at the national and European levels opposing stricter CO_2 emission standards, which supposedly specifically affect German car makers due to their focus on large high-margin vehicles [100]. Germany's aspiration to be the first country to pave the legal way to enable operation of automated vehicles on public roads is also linked to the relevance of the automotive sector.

9.3 Japan

Japan's automotive industry ranks among the nation's most significant industrial sectors. In 2019, the industry shipped 62.3 trillion JPY worth of product and accounted for some 18.8% of all Japanese manufacturing. In 1985, the corresponding figures were 27.7 trillion JPY and 10.4%, showing that, over the past 30 years, the automotive industry has increased in stature in Japan [52]. Large automobile manufacturers have developed together with the region where they are located—a so-called business castle town—and constitute an important part of the local economy. For example, Toyota Motor's headquarters is located in Toyota City, Aichi Prefecture, and its group companies, including parts suppliers, such as Toyota Industries Corp., Denso Corp., AISIN Corp., and Toyota Auto Body Co., Ltd., are dispersed throughout the neighboring cities.

The automotive industry forms part of a wide-reaching family of interrelated industrial sectors, together with associated industries such as manufacturing and procurement, sales, service, and shipping. At present, workers in automobile-related industries in Japan number 5.42 million, or 8.1% of Japan's total workforce [52]. The impressive breadth of the automotive industry's reach is also reflected in the high values of *power of dispersion* computed from input–output tables. Power of dispersion are economic indices used to quantify the relative importance of specific industrial sectors; the influence coefficient for a given sector is defined as the overall production impact—on the economy as a whole—induced by 1 unit of final demand in that sector, normalized such that the average industrial coefficient of all sectors takes the value 1. Based on data contained in the Updated *Input–Output Tables 2014* released by Japan's Ministry of Economy, Trade and Industry (METI), the three most influential subsectors of Japan's automotive industry are passenger vehicle, other vehicles and automobile components and accessories.

Although this analysis provides solid quantitative evidence for the centrality of the automotive industry in Japan, perhaps even more crucial is the industry's *symbolic* importance as the modern-day representative of Japan's industries, which manufactured sophisticated products requiring extensive teamwork—or, in the parlance of modern economics, products with *integral product architectures.* The term *product architecture*, as discussed by Takahiro Fujimoto at the Manufacturing Management Research Center (MMRC) (See e.g., Fujimoto [43]), refers to the design and optimization of the various interfaces through which the many components of complex modern products interact. In general, product architectures may be broadly classified into two categories: *modular* and *integral*. Products with *modular* architectures consist of individual components that are largely independent of each other; a good example of a modular-architecture product is a modern desktop computer. *Integral* architectures, by contrast, describe products with highly interdependent components that must all work together in complex ways for the product to perform its function—with automobiles furnishing a prime example. In products with integral architectures, changes to any single component require changes to surrounding components; thus, bringing products to completion requires extensive and intricate teamwork. Japan's talent for just this sort of teamwork is widely regarded as one of the key drivers of the nation's international competitiveness, and the automotive industry serves as a powerful symbol of Japan's unique strengths in this area.

9.4　Comparative Conclusions

Looking at the automotive industry's relevance for the national economy, Germany and Japan stand out from other industrial countries. For both countries, global leadership in the automotive sector is highly relevant for the prosperity of the respective national economy and its national innovation system, and hence of high priority for policymakers and parts of the general public. Hence, it is clear that for Germany and Japan vehicle automation is not only—and very likely not even primarily—a transport sector issue, but more a pursuit driven by industrial sector interests: maintaining the technological lead in this regard appears to be a question of national industrial reputation and prestige, as well as of one of economic importance in order to maintain the current advantage in the global competition in the automotive sector. This may apply even more to Germany with its focus on luxury cars and a sentiment that other countries' automotive industries managed to outperform the German car industry when it comes to vehicle electrification. In Japan, however, the state tends to directly intervene in scientific and technological innovation activities, because historically, scientific independence, or technological superiority, has been an important

Table 6 Key figures on travel demand and mode use in Germany and Japan [82, 98]

	Germany	Japan
Key figures on travel behavior (per person per day)		
Trip makers per day on weekdays (%)	86 (2017)	81 (2015)*
Trip makers per day on weekends (%)	77 (2017)	60 (2015)*
Trips per day, all persons (#)	3.1 (2017)	2.0 (2015)*
Daily distance, all persons (km)	39 (2017)	23 (2015)*
Daily travel time, all persons (min)	85 (2017)	56 (2015)*
Countrywide trip-based modal split on commuting trips		
On foot (%)	22 (2017)	8 (2010)
Bicycle (%)	11 (2017)	16 (2010)**
Personal Motorized Transport (PMT) driver (%)	57 (2017)	52 (2010)
Public transport (%)	10 (2017)	22 (2015)

* The values for Japan are averages for urban areas which in total account for more than 90% of Japan's population
** Includes motorcycles in addition to bicycles

concept in national (economic) security. In short: given the relevance of the automotive industry in both countries, political support for paving the way for vehicle automation can be expected, ranging from laying the legal foundations, to financial support for research and development (R&D) in this area.

10 Multimodal Transport Indicators

10.1 Comparative Statistics

See (Fig. 1).

10.2 Germany

The first German national household travel survey was conducted in 1976. For the first time, this dataset provided a systematic overview of travel behavior of German residents. In the four decades that followed, per capita travel grew considerably: While travel time per person per day increased by 25% from 68 to 85 min, daily travel distance increased by 44% from 27 to 39 km (excluding travel outside Germany). Much of this growth, and specifically the increase in travel speed which caused distances to grow more than travel times, can be attributed to increasing car ownership and use [126]. More and more people were able to afford a car, and consequently to

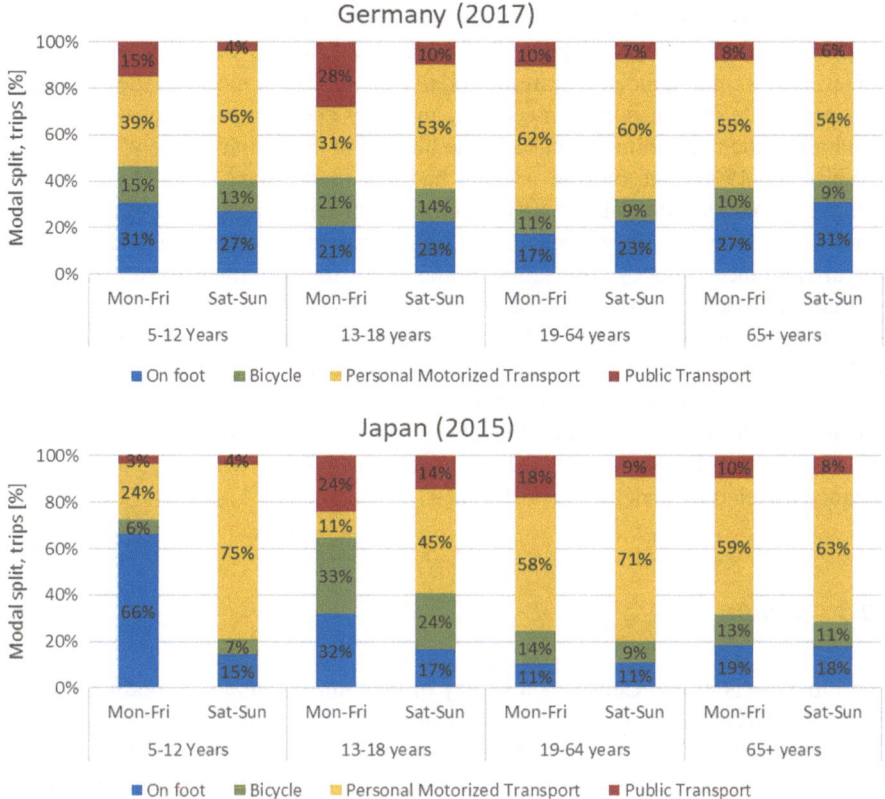

Fig. 1 Modal split (trips) in Germany (year 2017) and Japan (year 2015), distinguished by age and time of week (workday, weekend) (authors' analysis based on data from the National Travel Surveys)

travel longer distances. Currently, personal motor transport is used for about 57% of trips and about 75% of the daily mileage [99].

Overall, about a third of the German population uses a bicycle during the course of one week. Those who cycle use their bicycle on about a third of their trips, which translates to a national bicycle mode share of about 11%. The bicycle mode share is similar across the adult age groups and days of the week. Children and teenagers between 13–18 years show higher shares, with 21% of trips on weekdays and 14% of trips on weekends. The same applies to public transport, i.e., about a third of the population uses public transport during an average week, and if so, on about a third of their trips. This translates to a national public transport mode share of about 10%. Children and teenagers aged 13–18 years show the highest public transport modal split shares, using public transport for 28% of their trips on weekdays, i.e., mainly for commuting. Also, two-thirds of the population drive their cars during an average week, and drivers use their car on about two-thirds of their trips, yielding the national car driver mode share of about 43%, and car passenger mode shares of 14%. Adults between 19–64 years show the highest car usage mode shares (i.e., car driver and car passenger) of 62% on weekdays and 60% on weekends. Moreover, around 37% of Germany's inhabitants travel multimodally (as of 2017), meaning they use more than one mode of transport to meet their mobility needs in everyday travel. More than half of them combine the car and the bicycle in the course of the week [98].

Since the 1990s, the bicycle—which had lost mode share to motorized modes in the previous decades—has experienced a slight renaissance in Germany. Many German cities have actively promoted this development by expanding cycling infrastructure and increasing cycling safety. Some cities, for example Münster, Karlsruhe or Freiburg, reach cycling shares of more than 25% of all trips. In addition, electric bicycles (also dubbed "pedelecs") are becoming increasingly popular in Germany. In 2021, 13% of households owned at least one pedelec. Pedelecs enable greater distances to be covered due to the ride-assistance provided by the electric engine. The average distances traveled by bicycle and pedelec therefore also differ. In 2017, the average trip distance traveled by bicycle was 3.7 km, while the average trip distance traveled by pedelec was 6.1 km [98].

10.3 Japan

The Japanese National Census is a survey of the entire population conducted every five years, with the travel modes to and from work being surveyed every ten years since 1970. The motorization of Japan is said to have started around 1960, and the data shows an increase in car trips (from 15% in 1970 to 47% in 2010) and a decrease in the use of walking (from 23% in 1970 to 7% in 2010), railways (from 29% in 1970 to 25% in 2010) and buses (from 15% in 1970 to 7% in 2010). This data is limited to commuting to and from work for people aged 15 and over, and the actual situation of travel for personal purposes and on weekdays and weekends is discussed below.

Table 6 shows the travel modes share by age group, based on data from the National Person Trip Survey (NPTS) of Japan in 2015. The NPTS, which was conducted in 1987 for the first time, is a data set of one-day person trips during the autumn season for specific cities, selected in consideration of their size and relationship with surrounding cities, and obtained through a questionnaire survey. The target population was aged 5 years and over; 70 cities were covered by the NPTS survey in 2015.

The high weekday walking share of 66% for children aged 5–12 is due to the fact that walking to and from school is the designated mode of transport in many elementary schools in Japan. Many public elementary schools do not allow parents to drive their children to and from school. However, the private car trip share of 5–7-year-olds on weekends in 2015 was 75%; larger than the values for adults aged 19–64 (71%), and older people aged 65+ (63%). Japanese children tend to be transported in their parents' cars on weekends. The number of children killed in road traffic accidents in Japan has fallen sharply from 2115 in 1987 to 270 in 2015. This is thought to be due partly to the increase in children travelling by private car and a decrease in the number of children walking, rather than to safer towns. Among junior- and senior high school students aged 13–18, the share of public transport on weekdays is 24%, while cycling is as high as 33%. In Japan, many junior high school students are required to walk or cycle to school. However, the share of private cars is 45% on weekends.

The use of bicycles by children under 18 years in Japan for weekends and private purposes was 38% in 1987, but this figure had fallen sharply to 19% in 2015. Bicycling is recognized as a healthy and environmentally-friendly mode of transport, but this reduction may be partly due to the slow development of bicycle lanes in Japan. Although bicycles are categorized as vehicles in Japan, they are generally allowed on the sidewalk, and accidents between pedestrians and bicycles have occurred. Although the police authorities and the government have made it mandatory for new roads to have bicycle lanes and have emphasized that bicycles should be ridden on the road, not the sidewalk, it is not easy to break the Japanese custom, which has continued since the 1960s, of legally allowing bicycles on the sidewalk.

Among adults aged 19–64, the share of public transport on weekdays was 18%, and 9% on weekends, while the share of private cars was 58% on weekdays, and 71% on weekends, indicating a trend towards greater use of public transport for commuting on weekdays, and private cars for personal purposes on weekends. For the elderly aged 65 and over, the private car share was around 60% on both weekdays and weekends, indicating that there was little difference between weekdays and weekends. This would be because they are retired and not commuting. Looking at changes over time, the private car share of the elderly was 19% in 1987, but increased dramatically to 59% in 2015. On the other hand, the share of walking was 45% in 1987, but had fallen dramatically to 19% in 2015. This is thought to be due to motorization, which has created cities where it is difficult to live without a private car, and an increase in the number of healthy elderly people who own a driving license.

10.4 Comparative Conclusions

Current travel and mode use in Japan and Germany are a result of a long evolution driven by extending activity radiuses, i.e., longer trip distances, and a tendency to move towards faster and more convenient modes, mostly the private car, which today shows the highest mode share for both countries. It seems likely that the underlying trend towards more individualized destination choice—causing trip distances to grow and epitomized by the example of school choice—will continue as long as it is affordable for individuals. Throughout history, changes in the transport system, from increased travel mode choices to extensions of the infrastructure, have fueled this seemingly latent demand for longer trip distance at affordable generalized costs of travel. We do not see any reasons why vehicle automation may not follow this same path. The model-based analyses in chapters "Social Acceptance of CAD in Japan and Germany: Conceptual Issues and Empirical Insights and Transportation Effects of Connected and Automated Driving in Germany" illustrate on a quantitative

Table 7 Key figures on car ownership and use [4, 5, 7, 19, 51, 62, 63, 76, 77, 81, 87]

	Germany	Japan
Number of passenger cars on register per 1.000 population 1960	80 (1960)	5 (1960)
Number of passenger cars on register per 1.000 population 1990	420 (1990)	266 (1990)
Number of passenger cars on register per 1.000 population 2020	574 (2020)	490 (2020)
Passenger car average age (years)	9.8 (2020)	8.7 (2020)
Km per passenger car per year 1960 (km)	16,300 (1960)	19,080 (1960)
Km per passenger car per year 1990 (km)	15,300 (1990)	11,488 (1990)
Km per passenger car per year 2019 (km)	13,700 (2019)	8,860 (2019)
Passenger car occupancy rate (trip based)	1.3 (2019)	1.4 (2015)*

* Weighted average of the value for weekends and for holiday

basis that growth of overall travel demand is very likely a consequence of vehicle automation. The concrete incarnation of vehicle automation, however, will be key for the question which mode of travel—in essence individualized motorized modes or collective motorized modes—benefits the most from this development.

11 Vehicle Ownership and Use

11.1 Comparative Statistics

See (Table 7).

11.2 Germany

Between 1952 and 2019, the number of cars per capita in Germany increased more than 30-fold, and the number of kilometers per capita traveled by car about 12-fold. Until the 1970s, annual growth rates of car ownership and use were about 10%, and strong growth continued until about 1990 [19]. This was a period when economic growth (in West Germany) was strong, and the baby boomer generation entered adulthood and took up driving. However, also after 1990 both car ownership and car use continued to grow steadily, albeit on a somewhat lower level of about 1% annually.

In the past twenty years, there were about 3.5 million new passenger cars registered annually in Germany, while 3 million cars were deregistered and either scrapped or sold abroad [64]. This resulted in an annual net increase of the German passenger car stock of about half a million vehicles. Given the total number of cars on the register (about 40–50 million during that period), these figures translate to an annual growth rate of about 1%, and about one out of 15 vehicles being replaced every year. In other words, with this renewal rate it would take about 15–20 years to replace the entire existing car stock in Germany, exemplifying the inertia of the vehicle fleet.

The early growth of car ownership in Germany up until the 1990s was mainly driven by the economically active age groups acquiring their first vehicle. Step-by-step, this trend was replaced by two other developments which sustained the continued growth of car ownership from the 1990s: Firstly, a generation of seniors—specifically women—who were born before WWII and had lived without a driver's license and car were succeeded by a new generation of seniors who sustained driving habits which they had acquired during their younger years. This replacement of a senior generation without cars by auto-oriented seniors accounts for about a third of aggregate car-ownership growth since the turn of the millennium. Secondly, there is a trend to second and third cars in multi-driver households of all ages, i.e., there is a trend towards a personal car for each driver. This trend accounts for about two-thirds of the recent growth in car ownership in Germany [32].

There was a strong increase in car-sharing membership in Germany after the turn of the millennium, and especially after 2010 when automobile manufacturers such as Mercedes and BMW rolled out free-flow car-sharing schemes in various German cities [25]. This coincided with a decline of car-ownership rates among young adults in the 2000s. In the public and media debate, these indicators were interpreted as a sign for a new paradigm among the next generation of travelers, who were expected to turn to car usage instead of car ownership [67]. This trend may still materialize in the future, but long-term empirical evidence so far suggests different: Over the last decades, car ownership has grown more strongly than car use, leading to declining mileages per vehicle; there are more vehicles per licensed driver and also occupancy rates continue to decline (1950s: 2.5; 2000s: 1.5; 2019: 1.4, km-based occupancy rate) [19]. All of these are long-term indicators showing that the Germans in total continue to trend towards more individualized ownership and use of vehicles, which also conforms to other societal trends of continued individualization.

11.3   Japan

The number of automobiles owned in Japan, which in 1960 was just 1.4 million, grew to 58 million by 1990 and to 82 million by 2020, an increase of more than 60-fold in just over 60 years. These years also witnessed a dramatic evolution in the composition of Japan's automobile fleet: in 1966, passenger vehicles, freight vehicles, buses, and other vehicles (including two-wheeled vehicles) respectively accounted for 28.2%, 57.7%, 1.3%, and 12.8% of all Japanese vehicles, but by 1990 these shares had evolved to 56.8%, 36.1%, 0.4%, and 6.7%, and in 2020 they were 75.5%, 17.6%, 0.3%, and 6.7% [4]. Thus, we see that as the number of automobiles increased, their main use shifted from transporting goods to transporting people. Indeed, the number of passenger vehicles (including commercial vehicles) per 1,000 people exploded from 5 in 1960 to 266 in 1990, and then to 490 in 2020—demonstrating that notwithstanding the much-publicized aversion of young people to cars in recent years, the growth trend in the popularity of automobiles in Japan has remained undiminished.

Let us consider automobile ownership and usage in Japan on a household-by-household basis. According to "Passenger Car Market Trends in Japan" [53], an annual survey of households—including single-person households—conducted by the Japan Automobile Manufacturers Association (JAMA), 77.9% of all households in 2021 owned one or more passenger vehicles, with 34.5% of which owning two or more. According to the Automobile Inspection & Registration Information Association (AIRIA), the average number of private-use passenger vehicles owned per household was 1.04 in 2020, with lower numbers in major cities with extensive transit networks (such as Tokyo, with 0.42 vehicles per household, or Osaka, with 0.64), and higher numbers in rural areas (such as Fukui, Toyama, Yamagata, and Gunma prefectures, each of which had more than 1.6 vehicles per household) [6].

"Passenger Car Market Trends in Japan" also inquires how respondents use their most-recently purchased vehicles. Respondents (household's main driver) were asked to identify the primary purpose of their vehicle use. The two most common selections in 2021 were "Shopping, running errands, and other tasks" (42% of respondents), and "Commuting to work or school" (31%). Regarding the frequency of vehicle use, just under 40% of respondents reported using their cars seven days per week, while the average response was approximately five days per week.

In recent years, "Passenger Car Market Trends in Japan" has begun asking households owning 4-wheeled vehicles about the availability of car-rental or car-sharing services in their neighborhood, and how willing they might be to use such services. The fraction of respondents reporting the presence of a car-rental station within 10 min' walking distance of their homes was 35% in the Tokyo's 23 wards, averaged 31% in the 5 major cities (Tokyo's 23 wards, Yokohama, Kawasaki, Osaka, and Kyoto), and lower elsewhere in 2021. Similarly, the fraction of respondents within a 10-minute walk of a car-sharing station was 45% in the Tokyo's 23 wards, averaged 36% in the 5 major cities, and was lower elsewhere. However, when asked about current and past use, the share of households with experience of using was over 50% for car-rental services, but just 3% for car-sharing services. Similarly, in response to the question about intentions to use, only 16% of respondents answered in the affirmative for car sharing services.

While these statistics paint a quantitative picture of automobile ownership and use in Japan, we glean some cultural insight from the colloquial evolution of the Japanese-language phrase "three sacred treasures" (*sanshu no jingi*). Historically a reference to three priceless artifacts handed down through generations of Japanese imperial rulers as symbols of power, the phrase was adapted in the 1950s to describe three possessions then epitomizing the typical Japanese household's aspiration to a life of abundance: a black-and-white television, a washing machine, and a refrigerator. By the 1960s, economic progress had expanded expectations, and now the new three sacred treasures were a color television, an air conditioner—and a car. Indeed, until quite recently, ownership of a home and a private passenger vehicle were widely seen as the central material goals of Japanese families. The act of owning an asset entitles the owner to both residual profits from the asset and residual control rights over use of the asset, thus enhancing the asset's usefulness to the owner and creating powerful incentives to increase its value. At the same time, private asset ownership leads to increased energy usage and greater environmental impact due to production and disposal of assets, and creates inefficiency when assets lie idle and unused. Car sharing, enabled by autonomous vehicles and internet-of-things-based monitoring, may prove a powerful tool for alleviating these shortcomings while retaining the advantages of private ownership. Nonetheless, international comparative surveys conducted in 2017 show that Japanese consumers are far less willing than consumers in other nations to consider using ride-sharing services and other innovations of modern sharing economies—an attitude that seems as prevalent among younger Japanese consumers in their 20 s as among their older compatriots [75]. The growing popularity of car sharing in Europe and the US is far from guaranteed to be replicated in Japan.

11.4 Comparative Conclusions

As regards car ownership and use, there are substantial differences between the two countries: Germany has a higher level of car ownership and higher mileage per passenger car, leading to an overall higher level of driving per capita. However, the overall trends in the recent decades are similar: There has been an increasing trend towards individualized ownership and use of cars as exemplified by the increasing number of cars per household and declining occupancy rates. Both countries have also seen a public debate about a possible trend towards replacing private car-ownership by shared or rented vehicles. Even though transitioning to shared cars seems more logical in Japan given the low mileage of private cars there, car-sharing schemes appear to be more successful in Germany. However, neither in Germany nor in Japan have the car-rental or car-sharing schemes that have emerged in recent years substantially impacted the overall trend towards more individual ownership and use of cars. Automation of private vehicles is rather likely to reinforce prevailing trends toward higher car ownership because it increases the utility of private vehicles. Nevertheless, vehicle automation is being discussed as a possible game-changer because it could enable automated individual mobility services which would diminish the utility advantage that private vehicles have over the use of other modes. Chapter "Transportation Effects of Connected and Automated Driving in Germany" in this book analyzes likely impacts of automated service on car ownership in a model-based quantitative manner.

12 Road Transport Automation—Evolution and Policy Approach

12.1 Germany

Technological and social developments in information and communication technologies (ICT), transportation technologies and mobility are closely intertwined. Between the 1920 and 1950s, the motor car was adopted by the masses and became the dominant means of personal transport in Germany. This development created substantial pressure on governments to expand existing and create new road infrastructure, but also new challenges for public safety, transport efficiency and (later) environmental protection. The rise of semiconductor electronics, microprocessors and computers in the 1950 and 1960s apparently already offered options for technological fixes. Certain ICT components, such as loop detectors and ramp metering, bus automatic vehicle location or dynamic message signs, were added to the road transport system. However, the grand vision of centralized road traffic management using computers as a central planning and steering device had to be abandoned, and electronics systems were still too bulky, too expensive and not sufficiently reliable to become part of the average passenger car [130].

This situation had changed by the mid-1980s to early 1990s. After two energy price crises, with a growing environmental protection movement, increasing growth in vehicle miles traveled and approaching limits to further infrastructure expansion, safety and environmental concerns again became a focus of transportation policy. At the same time, ICT had become smaller, more powerful and cheaper, which led both industry and governments to rediscover and revisit their earlier ambitions. At this time, R&D on "artificial intelligence" enjoyed a new upswing, and ICT increasingly became subject to industrial policy disputes between the major national economies. As a consequence, specific application programs for ICT in important sectors were initiated. One of the central projects in this field in Germany in the 1980s was the EUREKA research program "PROMETHEUS" (1987–1994) [101, 102]. It was understood as an integrated transport concept in which the unwanted social and ecological consequences of individual transport were to be reduced, and at the same time its advantages further utilized by exploiting the problem-solving potential of new technologies, in particular by combining transport technology with ICT. The work carried out at that time, especially on system analysis and problem definition for the application of ICT in road traffic, had a significant influence on further R&D activities in Germany (and beyond), and probably still does today. This also applies to approaches for the development and implementation of automated driving within PROMETHEUS, which resulted in a number of demonstration vehicles being introduced in this period.

Around the middle of the 2000s—starting with the DARPA Grand Challenge and the private-sector activities it triggered—another "renaissance" of automated driving began [73]. This was due, among other things, to the fact that much more powerful and specifically more cost-effective components for automation hardware and software (especially machine learning methods) became available. In countries where globally important players of the vehicle and electronics industry or the so-called platform economy are based, these industries are considered to be key assets for the national economy, and enjoy direct and indirect political support. This was also the case in Germany. After initial activities like founding a working group on legal implications of automated driving at the Federal Highway Research Institute [44], an inter-ministerial working group on automated driving, and a large technology assessment scoping study (Ladenburg study, financed by Daimler and Benz Foundation [74]), the German government adopted its "Strategy for Automated and Connected Driving" in 2015 [20], and collaborated with the Automotive Industrial Association (VDA) in making the IAA 2015 a major showcase for its ambitions. Since then, a number of additional policy initiatives have been started. Federal and state research funding has been increased and partially redirected toward various—mainly technical and organizational—aspects of automated driving. In 2016, the Federal Minister of Transport and Digital Infrastructure had set up an "Ethics Commission on Automated and Connected Driving" with the task to reflect on a number of ethical and social aspects of automated driving [50]. The German road traffic law has been amended twice within the last five years—in 2017 in order to enable the integration of automated driving as a new principle into the regulatory framework for road traffic, and in 2021 in order to support the deployment of fully automated vehicles on German roads. This was accompanied by German activities on the international level, especially within the World Forum for Harmonization of Vehicle Regulations (UNECE WP.29). An Action Plan, "Research for autonomous driving", which further intensifies and coordinates the activities of three federal ministries in the field of connected and automated driving, was published in 2019 [22]. Beyond governmental actors and public research organizations, also industrial players like car manufacturers and Tier 1s have substantially increased their R&D activities and collaborated in a number of verification, validation and demonstration projects.

12.2 Japan

Automatic train operation (ATO) on railways has a long history and comprises two types: with or without a driver. The type in which a driver has boarded and completed safety checks has been tested since the 1960s. The first ATO test in Japan took place on the Nagoya Municipal Subway Higashiyama Line from 1960 to 1962. The first operational ATO with a driver was employed in the monorail providing transportation to the Japan World Exposition venue in 1970. Subsequently, ATO has been widely used mainly for railways such as subways and monorails, which do not intersect with other traffic horizontally, in Sapporo, Sendai, Tokyo metropolitan area, Kyoto, Osaka, Kobe, Fukuoka, and other cities. ATO without drivers is widely used in Automatic Guide Weight Transit (AGT), which is a medium-weight trucking system that automatically drives small, lightweight vehicles with rubber tires along the guideways of dedicated lanes. The Port Island Line of Kobe New Transit Co., Ltd., which opened in 1981, was the first practical AGT in Japan, and the world's first automatic unmanned operation system. Subsequently, AGT became commonplace throughout Japan.

The history of actual automated driving implementation on roads is not as extensive as that of railroads, although automated driving research on roads had already been initiated by the government in the early 1960s [122]. In addition to the difficulties associated with technological realization related to the complexity of the road environment, this lag resulted from different government agencies administering various laws governing automated driving on roads. The Road Transport Vehicle Act, which defines vehicle standards, and the Road Transport Act, which defines business requirements for buses, taxis, rental cars, and car-sharing, fall under the jurisdiction of the MLIT. The Road Traffic Act, which sets traffic rules, is under the jurisdiction of the National Police Agency (NPA). The Ministry of Internal Affairs and Communications (MIC) and Ministry of Economy, Trade and Industry (METI) are also involved in automated driving operations. To achieve the operation of automated driving on public roads, coordination among these ministries was required, which was challenging.

However, in the 2010s, the trend toward interagency collaboration accelerated. Since 2014, the Public–Private ITS Initiative Roadmap has been updated and published annually on the initiative of the Cabinet Secretariat and collaborations between ministries and agencies. The 2014 roadmap included an expected timeframe for Level 3 automated driving marketability (see the introduction for more information on Level 3). In the same year, the Strategic Innovation Promotion Program (SIP), a cross-ministry technology development project led by the Cabinet Office, was launched. These developments gradually laid the groundwork for cross-ministry discussions and instigated the cooperation between ministries and agencies necessary for the social implementation of automated driving. Under the SIP, a test was conducted in 2017 consisting of platooning trucks, as well as a test demonstration of an unmanned cart service centered on local roadside stations.

For the implementation of unmanned operation on public roads, operators must meet the requirements of the Road Traffic Act, under the jurisdiction of the NPA, and the Road Transport Vehicle Act, under the jurisdiction of the MLIT. Based on the Public–Private ITS Initiative Roadmap, two laws were amended in April 2020 to allow Level 3 automated driving operation on public roads under certain conditions. Thereafter, on March 4, 2021, Honda launched the world's first mass-market vehicle equipped with Level 3 functions. In addition, an amendment to the Road Traffic Act, which was passed in April 2022 and is expected to go into effect by the end of the same year, will allow Level 4 operation on public roads under certain conditions. The amendment defines automated driving without a driver in the vehicle as "specified automated driving" and requires permission from the Prefectural Public Safety Commission. The "Chief Supervisor of Specified Automatic Operation," who remotely monitors the vehicle, will have the same duties as the driver, such as providing first aid in the event of an accident.

12.3　Comparative Conclusions

Above all, this analysis on the evolution of road transport automation in Germany and Japan clarifies the long history of research and technology development dedicated to vehicle automation. This history goes back much further than the automation technology hype which has dominated the discussion since around 2010. While in Germany there was an early focus on road transport automation dedicated to the idea to render the use of road capacity more efficiently, transport automation technology in Japan was first mostly concentrated on rail transport and slowly spilled over to road transport. It is clear that in both countries, transportation automation technology development was fostered by the government and closely linked to the dominance of the transport and automotive industry. Laying the legal foundations quickly to enable real world operation of automated vehicles has obviously been an ambition in both countries: working towards this objective, both governments—in close interaction with the automotive industry—have amended relevant laws within the last five years. Given this temporal analogy, it appears that both governments intend to maintain, or even strengthen, the current competitive positions of the national automotive industries by permitting their national industrial players to test and deploy autonomous driving technologies on public roads before global processes of harmonization of vehicle regulations (like UNECE WP.29) have been concluded.

13　Conclusions and Outlook

This chapter has provided a broad overview of the framework conditions that characterize the transport systems in Germany and Japan today. It presented the respective historical evolution as well as selected key factors that shape contemporary passenger transport and are likely to be influential in the future. It is clear that vehicle automation will be embedded in such existing systems that are characterized by path dependencies and considerable degrees of inertia. With respect to the factors presented in this chapter, Germany and Japan partly differ substantially (e.g., with regards to car use) but partly share important similarities (e.g., with regard to the relevance of the automotive industry). Germany and Japan make suitable case studies because they represent prototypical paths of mobility development: Germany represents a typical European path which is relatively car-oriented but not as extreme as the North American transport development trajectory. Japan follows a mobility pathway which—relative to other large high-income countries—is somewhat less auto-oriented. Hence, the two countries possibly represent transport systems pathways that many emerging countries may be following.

At the same time, Germany and Japan stand out from other countries with regard to factors which may be influential in making them forerunners with regard to the implementation of vehicle automation technologies. Both may be motivated more than others to introduce automation technologies because their densely-populated and rapidly-aging countries may benefit more than others from vehicle automation. Addressing an emerging lack of drivers, sustaining the mobility of an aging population and harnessing the potential for increased space-efficiency of automated traffic are just examples. However, likely even more important—vehicle automation may be key to sustaining the technological advantage that Germany's and Japan's automotive industries have over automobile manufacturers from much of the rest of the world. Hence, there are strong economic and political motivations to lay the legal, infrastructural and technological groundwork for the introduction of vehicle automation.

This chapter conceptually discussed how the presented factors influence, firstly, the motivation to introduce vehicle automation; secondly, the speed with which vehicle automation is likely to diffuse in the coming decades; and thirdly, how autonomous vehicles may alter transport systems and mobility behavior in the coming decades. This chapter has therefore laid the groundwork for understanding the context for introducing vehicle automation, and provides a framework for interpreting the findings of the subsequent chapters in the overall context of the national transport systems. The subsequent chapters of this book focus on selected key issues in this context, using a range of methodological approaches.

References

1. ADAC. (2022). Benzin- und Dieselpreis: So entstehen die Spritpreise. Retrieved from https://www.adac.de/verkehr/tanken-kraftstoff-antrieb/tipps-zum-tanken/7-fragen-zum-ben zinpreis/.
2. Alexander Albrecht. (2021, 01.11.2021). So geht es mit den Hochstraßen weiter. *Rhein-Neckar-Zeitung*. Retrieved from https://www.rnz.de/region/metropolregion-mannheim_art ikel,-ludwigshafen-so-geht-es-mit-den-hochstrassen-weiter-_arid,764272.html.
3. Arbib, J., & Seba, T. (2017). *Rethinking Transportation 2020–2030, The Disruption of Transportation and the Collapse of the Internal-Combustion Vehicle and Oil Industries*. Retrieved from https://www.wsdot.wa.gov/publications/fulltext/ProjectDev/PSEPro gram/Disruption-of-Transportation.pdf
4. Automobile Inspection & Registration Information Association. (undated). The number of owned vehicles. Retrieved from https://www.airia.or.jp/publish/statistics/number.html.
5. Automobile Inspection & Registration Information Association (AIRIA). (2020a, Aug. 31, 2020). Average vehicle age. Retrieved from https://www.airia.or.jp/publish/file/r5c6pv000 000u7a6-att/02_syarei02.pdf.
6. Automobile Inspection & Registration Information Association (AIRIA). (2020b). Number of Private Vehicles per Household, News Release. Retrieved from https://www.airia.or.jp/pub lish/file/r5c6pv00000104ju-att/kenbetsu2022.pdf
7. Automobile Inspection & Registration Information Association (AIRIA). (undated, Aug. 31, 2020). Number of Vehicles Owned. Retrieved from https://www.airia.or.jp/publish/statistics/ number.html.
8. Bahamonde-Birke, F. J., Kickhofer, B., Heinrichs, D., & Kuhnimhof, T. (2018). A Systemic View on Autonomous Vehicles Policy Aspects for a Sustainable Transportation Planning. *DISP, 54*(3), 12–25. https://doi.org/10.1080/02513625.2018.1525197.
9. Balser, M. (2021). NGO-Bündnis fordert Ende der Auto-"Klüngelei". *Süddeutsche Zeitung*. Retrieved from https://www.sueddeutsche.de/wirtschaft/autoindustrie-lobbyismus-ngo-bun desregierung-1.5484277.

10. BBSR. (2017). *Raumordnungsbericht 2017. Daseinsvorsorge sichern*. Retrieved from Bonn: https://www.bbsr.bund.de/BBSR/DE/veroeffentlichungen/sonderveroeffentlichungen/2017/rob-2017.html.

11. Bormann, R., Bracher, T., Dümmler, O., Dünbier, L., Haag, M., Holzapfel, H., … Zoubek, H. (2010). *Neuordnung der Finanzierung des Öffentlichen Personennahverkehrs. Bündelung, Subsidiarität und Anreize für ein zukunftsfähiges Angebot*. Friedrich-Eber-Stiftung, Bonn.

12. Bormann, R., Fink, P., Holzapfel, H., Rammler, S., Thomas Sauter-Servaes, Tiemann, H., … Weirauch, B. (2018). *DIE ZUKUNFT DER DEUTSCHEN AUTOMOBILINDUSTRIE. Transformation by Disaster oder by Design?* Friedrich-Ebert-Stiftung, Bonn.

13. Bundesinstitut für Bevölkerungsforschung. (2022). Fakten. Lebendgeborene und rohe Geburtenziffer in Deutschland (1841–2019). Retrieved from https://www.bib.bund.de/DE/Fakten/Fakt/F01-Lebendgeborene-Geburtenziffer-ab-1841.html.

14. Bundesministerium der Finanzen. (2022). Zusammensetzung der Spritpreise. Retrieved from https://www.bundesfinanzministerium.de/Content/DE/Standardartikel/Themen/Steuern/2022-03-14-zusammensetzung-der-spritpreise.html.

15. Bundesministerium der Justiz. (2021a). *Bundesfernstraßengesetz (FStrG)*. Retrieved from https://www.gesetze-im-internet.de/fstrg/.

16. Bundesministerium der Justiz. (2021b). *Personenbeförderungsgesetz (PBefG)*. Retrieved from https://www.gesetze-im-internet.de/pbefg/BJNR002410961.html.

17. Bundesministerium der Justiz. (2021c). *Straßenverkehrs-Ordnung (StVO)*. Retrieved from https://www.gesetze-im-internet.de/stvo_2013/.

18. Bundesministerium der Justiz. (2021d). *Straßenverkehrsgesetz (StVG)*. Retrieved from https://www.gesetze-im-internet.de/stvg/.

19. Bundesministerium für Digitales und Verkehr (BMDV). (2022). Verkehr in Zahlen 2021/2022. Retrieved from https://www.bmvi.de/SharedDocs/DE/Artikel/G/verkehr-in-zahlen.html.

20. Bundesministerium für Verkehr und digitale Infrastruktur. (2015). *Strategie automatisiertes und vernetztes Fahren. Leitanbieter bleiben, Leitmarkt werden, Regelbetrieb einleiten*. Retrieved from https://bmdv.bund.de/SharedDocs/DG/Publikationen/DG/broschuere-strategie-automatisiertes-vernetztes-fahren.pdf?__blob=publicationFile

21. Bundesministerium für Verkehr und digitale Infrastruktur. (2016). *Bundesverkehrswegeplan 2030*. Retrieved from https://bmdv.bund.de/DE/Themen/Mobilitaet/Infrastrukturplanung-Investitionen/Bundesverkehrswegeplan-2030/bundesverkehrswegeplan-2030.html

22. Bundesregierung. (2019). *Aktionsplan Forschung für autonomes Fahren. Ein übergreifender Forschungsrahmen von BMBF, BMWi und BMVI*. Retrieved from https://www.bmbf.de/SharedDocs/Publikationen/de/bmbf/5/24688_Aktionsplan_Forschung_fuer_autonomes_Fahren.html

23. Bundesregierung. (2022). *Allgemeine Verwaltungsvorschrift zur Straßenverkehrs-Ordnung (VwV-StVO)*. Retrieved from https://www.verwaltungsvorschriften-im-internet.de/bsvwvbund_26012001_S3236420014.htm.

24. Bundestag. (2019). *Abgrenzung der Begriffe Schienenpersonenfern- und Schienenpersonennahverkehr sowie Zuständigkeiten für das Schienennetz*. Retrieved from https://www.bundestag.de/resource/blob/656022/4a8ace3f18d8824c5f92b30d81a887c0/WD-5-071-19-pdf-data.pdf.

25. Bundesverband Carsharing (bcs). (2021). *CarSharing 2021 in Deutschland*. Retrieved from https://www.carsharing.de/jahresberichte

26. Bureau of Transportation Statistics. (2022). World Motor Vehicle Production, Selected Countries. Retrieved from https://www.bts.gov/content/world-motor-vehicle-production-selected-countries.

27. BVG. (2022). Berlkönig Website. Retrieved from https://www.berlkoenig.de/.

28. Cabinet Office. (2020). FY2019 Annual Estimates of GDP. Retrieved from https://www.soumu.go.jp/main_content/000731590.pdf.

29. Central Intelligence Agency (undated). The world factbook. Country Comparisons Railways. Retrieved from https://www.cia.gov/the-world-factbook/field/railways/country-comparison.

30. Dargay, J., Gatley, D., & Sommer, M. (2007). Vehicle Ownership and Income Growth, Worldwide: 1960–2030. *Energy Journal*(28(4)), 143–170.

31. Destatis. (2017). Einkommens- und Verbrauchsstichprobe (EVS). Retrieved from https://www.destatis.de/DE/ZahlenFakten/GesellschaftStaat/EinkommenKonsumLebensbeding ungen/Methoden/Einkommens_Verbrauchsstichprobe.html.

32. Destatis. (2018). *Wirtschaftsrechnungen. Einkommens- und Verbrauchsstichprobe. Ausstattung privater Haushalte mit ausgewählten Gebrauchsgütern und Versicherungen.*

33. Destatis. (2021). *Volkswirtschaftliche Gesamtrechnungen. Bruttoinlandsprodukt, Bruttonationaleinkommen, Volkseinkommen. Lange Reihen ab 1925.*

34. Destatis. (2022a). Consumer Price index. Retrieved from https://www.destatis.de/EN/Themes/Economy/Prices/Consumer-Price-Index/_node.html.

35. Destatis. (2022b). Genesis-Online. Wirtschaftbereiche. Retrieved from https://www-genesis.destatis.de/genesis/online?operation=previous&levelindex=3&step=1&titel=Statistik+%28Tabellen%29&levelid=1655408708979&levelid=1655408687195#abreadcrumb.

36. Destatis. (2022c). Genesis Online. Gebiet, Bevölkerung, Arbeitsmarkt, Wahlen. Retrieved from https://www-genesis.destatis.de/genesis/online.

37. Destatis. (2022d). Regionaldatenbank Deutschland. Retrieved from https://www.regionalstat istik.de/genesis/online.

38. Destatis. (2022e). Wohnungsbestand nach Anzahl und Quadratmeter Wohnfläche. Retrieved from https://www.destatis.de/DE/Themen/Gesellschaft-Umwelt/Wohnen/Tabellen/wohnun gsbestand-deutschland.html.

39. Eisenmann, C., & Kuhnimhof, T. (2018). Some pay much but many don't: Vehicle TCO imputation in travel surveys. *Transportation Research Procedia, 32*, 421–435. https://doi.org/10.1016/j.trpro.2018.10.056.

40. Energy Information Center. (undated). Gas prices. Retrieved from https://pps-net.org/oilstand

41. Engartner, T. (2008). *Die Privatisierung der Deutschen Bahn. Über die Implementierung marktorientierter Verkehrspolitik.* Wiesbaden: VS Verlag für Sozialwissenschaften Wiesbaden.

42. FGSV (2006). Richtlinien für die Anlage von Stadtstraßen RASt06. FGSV Verlag, Köln.

43. Fujimoto, T. (2008). *Architecture-based Comparative Advantage in Japan and Asia.* Paper presented at the Manufacturing Systems and Technologies for the New Frontier, London.

44. Gasser, T. M., Arzt, C., Ayoubi, M., Bartels, A., Bürkle, L., Eier, J., … Vogt, W. (2012). *Rechtsfolgen zunehmender Fahrzeugautomatisierung. Gemeinsamer Schlussbericht der Projektgruppe.* Berichte der Bundesanstalt für Straßenwesen. Bergisch Gladbach.

45. Geospatial Information Authority of Japan (2016). National Area Survey. Area survey of prefectural cities, towns and villages nationwide. Retrieved from https://www.gsi.go.jp/KOK UJYOHO/MENCHO-title.htm.

46. Gil, P., & Schindler, C. (2020). Erfolgsfaktoren von Straßenbahnsystemen. Analyse der Straßen- und Stadtbahnsysteme in West- und Mitteleuropa. *Der Nahverkehr, 1/2020*(1/2020), 2–7.

47. Haefeli, U. (2016). Entwicklungslinien deutscher Verkehrspolitik im 19. und 20. Jahrhundert. In O. Schwedes, W. Canzler, & A. Knie (Eds.), *Handbuch Verkehrspolitik* (pp. 97–115). Wiesbaden: Springer Fachmedien Wiesbaden.

48. Hascher, M., & Zeilinger, S. (2001). Verkehrsgeschichte Deutschlands im 19. und 20. Jahrhundert. Verkehr auf Straßen, Schienen und Binnenwasserstraßen. Ein Literaturüberblick über die jüngsten Forschungen. In *Jahrbuch für Wirtschaftsgeschichte* (Vol. Vol. 2001.2001, pp. 18).

49. Hauptamt für Statistik und Wahlen des Magistrates von Groß-Berlin. (1947). *Berlin in Zahlen 1946–1947.* Retrieved from Berlin: https://digital.zlb.de/viewer/image/16320499_1946_1947/4/.

50. Bundesministerium für Verkehr und digitale Infrastruktur (2017). *Ethik-Kommission Automatisiertes und Vernetztes Fahren. Eingesetzt durch den Bundesminister für Verkehr und digitale Infrastruktur.*

51. Japan Automobile Manufacturers Association (JAMA). (2018). *World Motor Vehicle Statistics.*

52. Japan Automobile Manufacturers Association (JAMA). (2020). *The Motor Industry of Japan 2020*. Retrieved from https://www.jama.org/the-motor-industry-of-japan-2019-2/.
53. Japan Automobile Manufacturers Association (JAMA). (2021). *Passenger Car Market Trends in Japan* Retrieved from https://www.jama.or.jp/release/news_release/2022/1298/.
54. Japan Automobile Manufacturers Association (JAMA). (undated-a). Automobile related taxes. Retrieved from https://www.jama.or.jp/operation/tax/burden/index.html.
55. Japan Automobile Manufacturers Association (JAMA). (undated-b). User's burden of automobile taxes. Retrieved from https://www.jama.or.jp/operation/tax/burden/index.htm.
56. Japan Road Association. (2018). *History of Road Policy in Japan*. Retrieved from https://www.mlit.go.jp/road/road_e/pdf/ROAD2018web.pdf
57. JSCE. (2021). *Infura tairyoku shindan [Physical checkup of Japan's infrastructure (Road)]* (in Japanese). Retrieved from https://committees.jsce.or.jp/kikaku/system/files/2021CheckUP_ROAD.pdf.
58. Karl, A., Mehlert, C., & Werner, J. (2017). *Reformbedarf PBefG. Rechtsrahmen für Mobilitätsangebote mit flexibler Bedienung unter besonderer Berücksichtigung des Bedarfs in Räumen und für Zeiten mit schwacher Nachfrage.* KCW GmbH, Berlin.
59. Karl, A., Regling, L., Stein, A., & Werner, J. (2019). *PBefG Novelle: Zulassung App basierter Fahrdienste mit Augenmaß Thematischer Vorabauszug aus dem Gesamtbericht: Grundlagen für ein umweltorientiertes Recht der Personenbeförderung.* KCW GmbH, Berlin.
60. Katagi, J. (2012). *Nichi-doku Hikakukenkyū Shichōson Gappei-Heisei no Daigappei ha Naze Shinten Shitaka [Comparative Study of Municipal Mergers in Japan and Germany-Why did the great municipal mergers of the Heisei era progress?]*: Waseda University Press.
61. Kornhauser, A. (2013). *Uncongested Mobility for All, New Jersey's Area-wide aTaxi System.* Retrieved from Princeton University: http://orfe.princeton.edu/~alaink/NJ_aTaxiOrf467F12/ORF467F12aTaxiFinalReport.pdf.
62. Kraftfahrt-Bundesamt. (2021a). *Fahrzeugzulassungen (FZ) Bestand an Kraftfahrzeugen und Kraftfahrzeuganhängern nach Haltern, Wirtschaftszweigen 1. Januar 2021. FZ23.* Retrieved from Flensburg: https://www.kba.de/SharedDocs/Downloads/DE/Statistik/Fahrzeuge/FZ23/fz23_2021_pdf.pdf.
63. Kraftfahrt-Bundesamt. (2021b). *FFahrzeugzulassungen (FZ) Bestand an Kraftfahrzeugen und Kraftfahrzeuganhängern nach Fahrzeugalter 1. Januar 2021. FZ 15.* Retrieved from Flensburg: https://www.kba.de/SharedDocs/Downloads/DE/Statistik/Fahrzeuge/FZ15/fz15_2021.pdf?__blob=publicationFile&v=2.
64. Kraftfahrtbundesamt. (2020). *Fahrzeugzulassungen (FZ). Neuzulassungen von Kraftfahrzeugen und Kraftfahrzeuganhängern-Monatsergebnisse. Dezember 2020.* Retrieved from Flensburg: https://www.kba.de/SharedDocs/Publikationen/DE/Statistik/Fahrzeuge/FZ/2020_monatlich/FZ8/fz8_202012_pdf.pdf?__blob=publicationFile&v=10.
65. Krummheuer, F. (2014). *Marktöffnung bei kommunalen Bahnen. Metros, Stadt- und Straßenbahnen im Wettbewerb.* Wiesbaden: Springer VS Wiesbaden.
66. Kuhnimhof, T., & Eisenmann, C. (2021). Mobility-on-demand pricing versus private vehicle TCO: how cost structures hinder the dethroning of the car. *Transportation.* https://doi.org/10.1007/s11116-021-10258-5.
67. Kuhnimhof, T., & Le Vine, S. (2021). Next Generation Travel: Young Adults' Travel Patterns. In R. Vickerman (Ed.), *International Encyclopedia of Transportation* (pp. 215–221). Oxford: Elsevier.
68. Kuhnimhof, T., Rohr, C., Ecola, L., & Zmud, J. (2014). Automobility in Brazil, Russia, India, and China-Quo Vadis? *Transportation Research Record: Journal of the Transportation Research Board, 2451*, 9.
69. Ladd, B. (2008). *Autophobia: Love and hate in the automotive age.* Chicago: University of Chicago Press.
70. Lawrence D. Burns, W. C. J., Bonnie A. Scarborough. (2013). *Transforming Personal Mobility.* The Earth Institute, Columbia University, New York.
71. Lincoln Institute of Land Policy. (2016). The Atlas of Urban Expansion. Retrieved from http://www.lincolninst.edu/subcenters/atlas-urban-expansion/documents/table-urban-land-cover-data.xls.

72. Matsuura, S. (1998). A Study on the Road works before the World War in Relation to the Policy of Road Works. *Historical studies in civil engineering, 18.*
73. Matthaei, R., Reschka, A., Rieken, J., Dierkes, F., Ulbrich, S., Winkle, T., & Maurer, M. (2015). Autonomes Fahren. In H. Winner, S. Hakuli, F. Lotz, & C. Singer (Eds.), *Handbuch Fahrerassistenzsysteme: Grundlagen, Komponenten und Systeme für aktive Sicherheit und Komfort* (pp. 1139–1165). Wiesbaden: Springer Fachmedien Wiesbaden.
74. Maurer, M., Gerdes, J. C., Lenz, B., & Winner, H. (2016). *Autonomous Driving.*
75. Ministry of Internal Affairs and Communications (MIC). (2017). *Present and Future of Smartphone Economy, 2017.* Retrieved from https://www.soumu.go.jp/johotsusintokei/linkdata/h29_01_houkoku.pdf.
76. Ministry of Internal Affairs and Communications (MIC). (undated-a). Historical Statistics of Japan. Retrieved from https://warp.da.ndl.go.jp/info:ndljp/pid/11423429/www.stat.go.jp/english/data/chouki/index.html.
77. Ministry of Internal Affairs and Communications (MIC). (undated-b). Population Estimates. Retrieved from https://www.e-stat.go.jp/statistics/00200524.
78. Ministry of Internal Affairs and Communications in Japan. (undated). Shichōson Sū no Henkan to Meiji Shōwa no Daigappei no Tokuchō [Changes in the Number of Municipalities and Characteristics of great municipal mergers of Meiji and Showa Era]. Retrieved from https://www.soumu.go.jp/gapei/gapei2.html.
79. Ministry of Internal Affairs and Communications in Japan (MIC). (2019). Annual Report on the Family Income and Expenditure Survey 2019. Retrieved from https://www.e-stat.go.jp/statistics/00200561.
80. Ministry of Land Infrastructure Transport and Tourism (MLIT). (2012). *WHITE PAPER ON LAND, INFRASTRUCTURE, TRANSPORT AND TOURISM IN JAPAN, 2012.* Retrieved from https://www.mlit.go.jp/english/white-paper/2012.pdf.
81. Ministry of Land Infrastructure Transport and Tourism (MLIT). (2015). *Road Traffic Census 2015* (in Japanese). Retrieved from https://www.e-stat.go.jp/statistics/00600580.
82. Ministry of Land Infrastructure Transport and Tourism (MLIT). (2016). Number of People Going Out Lowest Since Survey Began: Release of the Preliminary 2015 National Urban Transportation Characteristics Survey. Retrieved from https://www.mlit.go.jp/common/001156133.pdf.
83. Ministry of Land Infrastructure Transport and Tourism (MLIT). (2017). Kokutetsu no bunkatsu min-eika kara 30nen wo mukaete [On the Occasion of the 30th Anniversary of the Division and Privatization of Japan National Railways]. *Transportation & economy 77*(7), 9. Retrieved from https://www.mlit.go.jp/common/001242868.pdf.
84. Ministry of Land Infrastructure Transport and Tourism (MLIT). (2018a). Railway Statistics Annual Report 2018. Retrieved from https://www.mlit.go.jp/tetudo/tetudo_tk2_000051.html.
85. Ministry of Land Infrastructure Transport and Tourism (MLIT). (2018b). *Roads in Japan.* Retrieved from https://www.mlit.go.jp/road/road_e/pdf/ROAD2018web.pdf.
86. Ministry of Land Infrastructure Transport and Tourism (MLIT). (2019a). *Current Status and Issues of the Regional Public Transport, 2019* Retrieved from https://www.mlit.go.jp/policy/shingikai/content/001311082.pdf.
87. Ministry of Land Infrastructure Transport and Tourism (MLIT). (2019b). Survey on Motor Vehicle Fuel Consumption. Retrieved from https://www.e-stat.go.jp/statistics/00600370.
88. Ministry of Land Infrastructure Transport and Tourism (MLIT). (2020). *2040 Vision for Roads in Japan. To shape a better future for people.* Retrieved from https://www.mlit.go.jp/road/vision/index.html.
89. Ministry of Land Infrastructure Transport and Tourism (MLIT). (2021). *Roads in Japan.* Retrieved from https://www.mlit.go.jp/road/road_e/pdf/ROAD2021web.pdf.
90. Ministry of Land Infrastructure Transport and Tourism (MLIT). (undated-a). Asia Highway. Retrieved from https://www.mlit.go.jp/kokusai/kokusai_tk3_000071.html.
91. Ministry of Land Infrastructure Transport and Tourism (MLIT). (undated-b). https://wwwtb.mlit.go.jp/tohoku/content/000181559.pdf. Retrieved from https://wwwtb.mlit.go.jp/tohoku/content/000181559.pdf.

92. Ministry of Land Infrastructure Transport and Tourism (MLIT). (undated-c). *Law on Partial Amendment to the Law on Revitalisation and Revitalisation of Regional Public Transport*. Retrieved from https://www.city.yokohama.lg.jp/kurashi/machizukuri-kankyo/kotsu/tos hikotsu/seisaku/kako/past.files/0167_20210714.pdf.

93. Miriam Dross, Nadja Salzborn, Katrin Dziekan, & Klinski, S. (2021). *Klimaschutzinstrumente im Verkehr. Damit das Recht dem Klimaschutz nicht im Weg steht Vorschläge zur Beseitigung von Hemmnissen im Straßenverkehrsrecht*. Retrieved from Dessau-Rosslau: https://www. umweltbundesamt.de/dokument/damit-das-recht-dem-klimaschutz-nicht-im-weg-steht.

94. MOIA. (2022). MOIA Website. Retrieved from https://www.moia.io/en.

95. National Institute of Population and Social Security Research (IPSS). (2017). *Population Projections for Japan: 2016–2065*. Retrieved from Tokyo: https://www.ipss.go.jp/pp-zen koku/j/zenkoku2017/pp29_ReportALL.pdf.

96. National Institute of Population and Social Security Research (IPSS). (2018). *Household Projections for Japan:2015–2040*. Retrieved from https://www.ipss.go.jp/pp-ajsetai/j/HPR J2018/t-page.asp.

97. Nihon Bus Association. (2020). Japanese Bus Business. Retrieved from https://www.bus.or. jp/about/pdf/2020_busjigyo.pdf.

98. Nobis, C., & Kuhnimhof, T. (2018). *Mobilität in Deutschland-MiD. Ergebnisbericht*. Retrieved from https://www.bmvi.de/SharedDocs/DE/Anlage/G/mid-ergebnisbericht.pdf?__ blob=publicationFile.

99. Nobis, C., Kuhnimhof, T., Follmer, R., & Bäumer, M. (2019). *Mobilität in Deutschland−MiD: Zeitreihenbericht 2002−2008−2017. Studie von infas, DLR, IVT und infas 360 im Auftrag des Bundesministeriums für Verkehr und digitale Infrastruktur (FE-Nr. 70.904/15)*. Retrieved from Bonn, Berlin: http://www.mobilitaet-in-deutschland.de/pdf/MiD2017_Zeitreihenber icht_2002_2008_2017.pdf.

100. Nowack, F., & Sternkopf, B. (2015). *Lobbyismus in der Verkehrspolitik. Auswirkungen der Interessenvertretung auf nationaler und europäischer Ebene vor dem Hintergrund einer nachhaltigen Verkehrsentwicklung*. IVP-Discussion Paper, No. 2015 (2), Technische Universität Berlin, Fachgebiet Integrierte Verkehrsplanung, Berlin

101. Panik, F. (1987). *Automobiltechnik als Korrektiv menschlichen Unvermögens? FhG-Berichte 4–87*, 25–29

102. Prätorius, G. (1993). *Das PROMETHEUS-Projekt. Technikentstehung als sozialer Prozeß*. Wiesbaden: Gabler Verlag.

103. Reinhardt, W. (2018). *Öffentlicher Personennahverkehr*. Springer Vieweg, Wiesbaden.

104. Sack, D. (2007). Mehrebenenregieren in der europäischen Verkehrspolitik. In O. Schöller, W. Canzler, & A. Knie (Eds.), *Handbuch Verkehrspolitik* (pp. 176–199). Wiesbaden: VS Verlag für Sozialwissenschaften.

105. Saito, T. (1997). *A social history of vehicles, from the rickshaw to the automobile*: Tyuou Kousya.

106. Schönharting, J., & Schuhmann, M. (2010). Die Entwicklung der Verkehrsplanung bis heute. *Straßenverkehrstechnik, 54*(5).

107. Seum, S., Schulz, A., & Kuhnimhof, T. (2020). The evolutionary path of automobility in BRICS countries. *Journal of Transport Geography, 85*. https://doi.org/10.1016/j.jtrangeo. 2020.102739.

108. Shimada, A. (2018). "Heisei no Daigappei" no Sōkatsuteki Kentō [A Comprehensive Study of the "the great municipal mergers of the Heisei era"]. *Chihōjichi Fukuoka, 64*, 34. https:// doi.org/10.32232/chihoujichifukuoka.64.0_3.

109. Sony Assurance Inc. (2021). Zenkoku Car life jittai Chose [National Survey on Car life 2021. Retrieved from https://from.sonysonpo.co.jp/topics/pr/2021/08/20210824_01.html.

110. Stadt Aachen. (2016). *Statistisches Jahrbuch. Bevölkerungszahlen, Arbeitsmarktdaten und vieles mehr*. Retrieved from Aachen: https://www.aachen.de/DE/stadt_buerger/pdfs_stadtbu erger/pdf_statistik/statistisches_jahrbuch_2015.pdf.

111. Stadt Aachen. (2022a). Altstadtquartier Büchel. Retrieved from https://buechel-aachen.de/ was-bisher-geschah/.

112. Stadt Aachen. (2022b). *Stellplatzsatzung der Stadt Aachen*. Retrieved from https://www.aac hen.de/de/stadt_buerger/politik_verwaltung/stadtrecht/pdfs_stadtrecht/stellplatzsatzung.pdf
113. Statistics Bureau & Ministry of Internal Affairs and Communications (2015). 2015 Census. Retrieved from https://www.stat.go.jp/data/kokusei/2015/.
114. Statistics Bureau & Ministry of Internal Affairs and Communications. (2020). *2020 Population Census*. Retrieved from https://www.stat.go.jp/data/kokusei/2020/kekka.html.
115. Statistics Bureau of Japan. (2022a). Annual Report on Internal Migration in Japan Derived from the Basic Resident Registration 2022 Retrieved from https://www.stat.go.jp/data/idou/.
116. Statistics Bureau of Japan. (2022b). Japan Statistical Yearbook 2022. Retrieved from https://www.stat.go.jp/data/nenkan/index1.html.
117. Takano, Y., & Taniguchi, M. (2018). A Study on the government expenditure for public transportation policy by municipalities-An analysis of the questionnaire survey targeting all cities in Japan - (in Japanese). *Journal of the City Planning Institute of Japan, 53*(3), 7.
118. Takebe, K. (2003). *Road*. Shuppankyoku: Hosei University.
119. The Global Economy. (2019). Roads quality-Country rankings. Retrieved from https://www.theglobaleconomy.com/rankings/roads_quality/.
120. Tilmann, C. (2010, 16.10.2010). Berlin, wie es nie gebaut wurde. *Der Tagesspiegel*. Retrieved from https://www.tagesspiegel.de/kultur/architekturgeschichte-berlin-wie-es-nie-gebaut-wurde/1885378.html.
121. Tottori Prefecture. (undated). Prefecture Tottori WEB site: Special topic of National population census (in Japanese). Retrieved from https://www.pref.tottori.lg.jp/288728.htm.
122. Tsugawa, S. (2015). Jidōūnten Shisutemu no 60nen [Sixty Years of Automated Driving Systems]. *Journal of The Society of Instrument and Control Engineers, 54*(11), 5. https://doi.org/10.11499/sicejl.54.797.
123. VDV. (2016). Finanzierung. Retrieved from https://www.mobi-wissen.de/Finanzierung/Fin anzierung#:~:text=Die%20Finanzierung%20des%20%C3%96ffentlichen%20Personenna hverkehrs,%C3%B6ffentlichen%20Hand%20an%20die%20Verkehrsunternehmen.
124. Verkehrsverbund Berlin-Brandenburg. (2017). *Zahlen und Fakten 2017*. Retrieved from Berlin: https://www.vbb.de/presse/publikationen/zahlen-fakten.
125. Weigelt, H. (2010). Von den Anfängen der Eisenbahn bis zur Glanzzeit deutscher Länderbahnen-ein Überblick. *Eisenbahntechnische Rundschau*(12/2010), 9. Retrieved from https://www.forschungsinformationssystem.de/servlet/is/349815/.
126. Weiß, C., Chlond, B., Behren, S. v., Hilgert, T., & Vortisch, P. (2016). *Deutsches Mobilitätspanel (MOP)-Wissenschaftliche Begleitung und Auswertungen Bericht 2015/2016: Alltagsmobilität und Fahrleistung*. Institut für Verkehrsforschung, Karlsruhe.
127. Wisotzky, K., & Josten, M. (2018). *Essen. Geschichte einer Großstadt im 20. Jahrhundert*. Münster: Aschendorff.
128. World Bank. (2022). World Development Indicators. Retrieved from https://databank.worldb ank.org/source/world-development-indicators.
129. Wüstenberg, D. (2021). Änderungen im Personenbeförderungsgesetz 2021. *RdTW Recht der Transportwirtschaft, 9*(7), 9.
130. Zimmer-Merkle, S., Fleischer, T., & Schippl, J. (2021). 1.4 Mobilität und Technikfolgenabschätzung: zwischen Mobilitätsermöglichung und Mobilitätsfolgenbewältigung. In S. Böschen, A. Grunwald, B.-J. M. A. Krings, & C. Rösch (Eds.), *Technikfolgenabschätzung: Handbuch für Wissenschaft und Praxis* (1 ed., pp. 97–112). Baden-Baden: Nomos Verlagsgesellschaft mbH & Co. KG.

Governance, Policy and Regulation in the Field of Automated Driving: A Focus on Japan and Germany

Yukari Yamasaki, Torsten Fleischer, and Jens Schippl

Abstract Sociotechnical development is often described as an evolutionary process of a series of connected changes in different domains, including technology, the economy, institutions, innovation policies, behavior, culture, ecology, and belief systems. Many experts point to the great transformative potential of automated vehicles in the mobility sector, and to a variety of pathways that lead to an imagined future made possible by automated driving technologies. However, the differences in state behavior and governance approaches which are entangled in such emerging technologies are less understood, despite their potential to influence the trajectory of sociotechnical development. This chapter examines the modes and methods of governance in Japan with respect to automated driving. In order to illuminate the Japanese characteristics, we compare them with the German approach. We provide a brief comparison of the two democratic and capitalist countries from three perspectives—politics, polity, and policies. We then present Japan's policy process, policy actors, and recent changes in its approach to automated driving. In Japan, automated driving is interrelated with other policy areas, such as science and technology, information technology, and demographic change issues, and has been contextualized primarily in relation to the economy, particularly during the term of Prime Minister Shinzo Abe (2012–2020). The state has historically tended to intervene in technological development, and in recent years the Cabinet has attempted to exercise a more top-down political leadership through policy conferences. While letting the government appear to be taking a leadership role, relevant industrial players also seem to exert a significant influence on the direction of automated driving policies through both formal and informal channels. To enable effective and efficient governance in complex fields such as mobility, researchers, policymakers, and others involved in

Y. Yamasaki (✉) · T. Fleischer · J. Schippl
Institute for Technology Assessment (ITAS), Karlsruhe Institute of Technology (KIT), Karlsruhe, Germany
e-mail: yukari.yamasaki@kit.edu

T. Fleischer
e-mail: torsten.fleischer@kit.edu

J. Schippl
e-mail: jens.schippl@kit.edu

© The Author(s) 2025
C. Eisenmann et al. (eds.), *Acceptance and Diffusion of Connected and Automated Driving in Japan and Germany*, https://doi.org/10.1007/978-3-031-59876-0_3

governance require a good understanding of the factors that could influence future development pathways. Future research should build on these findings and conduct further comparative analyses between countries that have the potential to play a leading role in the implementation of automated driving.

1 Introduction

Automated driving (AD) has long been imagined and researched by governments. In this chapter, Japan is the center of interest. At relevant points, a brief comparison is made with the situation in Germany. In the early 1960s, a national research institute in Japan conducted research on automated highway systems [54]. Yet, since the early 2010s, driven by technical progress and unsolved challenges in the mobility sector, interest in the development of this technology has increased considerably in both the Japanese and German governments. This has been linked with growing expectations that AD will be a potential solution to various societal problems, and that pressure from non-traditional companies may change the player network of the domestic automotive industry. Accordingly, increasing budgets have been allocated to the research and development (R&D) of AD and its social implementation. Many experts point to the great transformative potential of automated vehicles for the mobility sector, and to a variety of different pathways that lead to an imagined possible future, with the associated need for governance [44].

For the two governments, transportation systems and mobility issues have been a subject of public policy from the perspective that they are an important part of human life and constitute the basic infrastructure necessary for many aspects of economic, social, and political activities. Along with the great gains in prosperity and quality of life through improved transport, however, the governments are also required to address the negative impacts of the externalities of mass car use, such as congestion, accidents, poor air quality, physical severance, social exclusion, and inactivity [7]. In addition, automobile policy has been a part of the industrial policy in terms of its significance as a key domestic industry in both countries. In recent years, new technologies and services in mobility fields, such as AD, have been addressed more frequently in science, technology and innovation policies. Thus, AD is discussed in different policy areas depending on how AD—technology, technology-applied products and systems, innovation, or means of transport—is conceptualized and contextualized in policy areas by experts and policymakers. In turn, different measures are discussed and/or implemented, such as transport planning and service provision, laws and regulations in social implementation, and subsidies and investment in R&D and infrastructure.

Different countries have different sociotechnical development paths. Even if it is recognized that we are in "a once-in-a-century period of profound transformation" of the automobile industry, as Toyota Motor Corporation's Annual Report 2018 states, technological innovation alone does not explain or predict how the sociotechnical system will develop in a country. The transitions are the result of a set of connected changes in different domains, such as technology, the economy, institutions, innovation policies, behavior, culture, ecology, and belief systems [40, 53]. The main objective of this chapter is to achieve a better understanding of the role and relevance of political institutions in the development of automated vehicles in Japan, and on a more general level in Germany. The idea is to focus on state actions anchored in each country's history, culture, and ideology. AD is integrated into various policy areas, and the state can have different policy strategies and measures, which in part shapes the ongoing mobility transition. How the two states have actually been involved varies according to the ideological lens through which the state is viewed—neoliberal market orientation or 'welfare model,' or a blend of both [7]. Affected by this, differences in current sociotechnical systems manifest themselves as car-dependent societies or societies with well-developed public transportation systems. While it is not possible to fully anticipate the mobility transition due to the complexity of interwoven domains, an exploration of political institutions and governance styles in AD will deepen our understanding of the different state approaches and their potential influence on sociotechnical development pathways. The core of this chapter is a detailed analysis of the Japanese situation, strongly based on relevant policy documents in the field of AD. To clarify the characteristics in Japan, we compare the Japanese situation with that in Germany on several central points. Germany is well-suited as a contrast to Japan, as both countries have a globally leading automotive industry and are dependent on technical progress to maintain this leading role. Therefore, AD is a mandatory topic in both countries, not only for industry but also for innovation policy.

Section 2 compares some characteristics of state actions in Japan with those of Germany in three aspects—politics, polity, and policies. While explaining the role of government as shaped by traditionally observed actors' relationships and expectations in different policy topics, such as industry, science and technology, and public transportation, we explore the three aspects of state actions in both countries with regard to AD. Based on the characteristics discussed in Sect. 2, Sect. 3 examines the situation in Japan. The results of an empirical study on policy processes, policy actors, and recent changes in their approaches surrounding the topic of AD is described, with a particular focus on the period of the late former Prime Minister Shinzo Abe (2012–2020). In the final section, we reflect on some of the key findings from the study and consider how different political institutional settings and governance approaches may affect the trajectory of sociotechnical development.

2 Different Governance Styles in Democratic Capitalist Countries

The modes and methods of governance of AD vary from country to country, influencing the trajectory of sociotechnical development. Although Japan and Germany are both described as democratic and capitalist countries, their institutional patterns of political and economic governance differ. In this section, we describe the differences in state approaches to AD in Japan and Germany from the perspectives of politics, polity, and policies, considering the historical context of each country. Deconstructing state action into the three components helps to analyze complex state approaches in the governance of AD [45].

2.1 Politics: Actor Relationships and Interactions Influenced by (Expected) Roles

2.1.1 State-Business Relations in Capitalist Countries

Politics of interactions and power relations, as well as discursive interactions among political actors and communicative discourse to the public, have a profound impact on policy trajectories [45]. For a better understanding of the role of state and state-firm relations, we introduce the classifications of capitalism by Schmidt [45]. Based on the simplest dichotomy of capitalism—liberal market economies (LMEs) and coordinated market economies (CMEs)—by Hall and Soskice [15], Schmidt offered a third variety that can apply to countries that do not fit the binary division, namely, state-influenced market economies (SMEs). In the first variety of LMEs, the state provides a high degree of autonomy to economic agents in market capitalism and acts as an arbiter. The enabling state in the second variety of CMEs encourages associational governance and negotiation among private agents in managed capitalism, and acts as a facilitator. In the third variety of state-influenced market economies (SMEs), the interventionist state directly coordinates and intervenes in private activities in state-enhanced capitalism, and acts as a leader [45].

Looking at state-firm relations and the expected role of the state, especially in economy, industry, and science and technology policy, Germany may count as a CME, while Japan is more of an SME. Yet their behavior may shift slightly to other capitalism types depending on the administration and policy areas at a particular time, and hence, they cannot automatically be placed in the respective categories. Historically, however, the focus of post-war political economy design was different in each country: for Japan, to catch up with and surpass the West through national planning and industrial strategy and social integration through growth; and in West Germany, the consensus-oriented construction and protection of national social cohesion and solidarity with the principle of social equilibrium (Yamamura and Streeck [60], p. 2, Hundt and Uttam [17]). Despite a trend toward more "neo-liberalism" and associated privatization in some sectors in both countries since the 1980s, this did not result in a complete slide from *faire* to *laissez-faire* by leaving everything up to market actors in all public services and industries, because political actors never thoroughly embraced the market-oriented philosophy [33]. In Japan, political actions based on a "neoliberal" mindset were adaptive while maintaining the characteristics of the developmental state, rather than transformative [58]; the state has believed in the need to take initiatives in the economy in order to control the results of competition in a particular direction, while industry also expects state leadership to set the overall goal [59]. The primary goal has consistently been economic development, not the establishment of a liberal market economy, and deregulation and economic openness approaches have been justified by the need to increase the international competitiveness of Japanese companies and industries, or the economy as a whole (Anchordoguy [1], chapter "Business Analysis and Prognosis Regarding the Shared Autonomous Vehicle Market in Germany", Hundt and Uttam [17]). In this sense, privatization and (de)regulation are merely one of the many policy tools to achieve the goal, which will be adopted when it is considered to contribute to its achievement [43]. As the state actively seeks a way to coordinate and develop the economy as a whole using different tools, the approach is different from *laissez-faire*.

Germany probably maintains its character as a CME in economy, industry and science and technology policy. However, Japan can still be included in the category of SMEs, given the traditionally larger role of the state, which has remained prominent despite significant retreat, as well as the close connection between the state and business, and the paternalistic firm-labor relationship [45]. What both countries have in common is a dense web of interrelationships between firms, their business partners, and government agencies. Industry players, especially automakers, appear to have exerted a significant influence on policy direction, particularly automotive policy, through established channels. However, a difference may be observed in that the Japanese government shows an outwardly more direct attitude toward orchestrating and guiding the private sector, whereas the German government rather facilitates coordination of the private sector [55]. This does not imply that, in Japan, the government engages in tyranny in the name of economic development, nor does it negate pluralist arguments that emphasize the role of non-state actors, such as industry, finance, and interest groups (for an overview of previous literature, see Mogaki [33]). The close relationship between actors bounded by socially-embedded communal norms, described as Japan's collective capitalism [17] or communitarian capitalism [1], has enabled the prominent, if not dominant, role of the state. Consequently, compared to LMEs and CMEs, the Japanese state has taken more than a facilitatory role by directly influencing the national trajectory in the economy, as well as science, technology and innovation. Following the so-called "lost decades"[1] due to economic stagnation, the state believes in the need for its traditional leadership role to initiate an economic revival, and is looking to rely on large domestic industrial powers such as Toyota Motor Corporation.

The relationship between the state and business sector depends on whether it is a public transportation operator or a manufacturer developing mobility-related technologies. For example, in Japan, after the deregulation of the taxi industry since the 1990s, operators lobbied the government to reinstate regulations due to

---

[1] Japan's "lost decade" refers to approximately 10 years since 1991, when the bubble economy collapsed. In the following decades, GDP growth remained sluggish affected by the global economic recession due to the global financial crisis (in 2008) and the COVID-19 pandemic (since 2020), as well as the Great East Japan Earthquake (in 2011), so that the years from 1990 to the present are sometimes collectively referred to as the "lost decades".

intensified competition, and regulations were reintroduced to protect taxi drivers. Likewise, the industry can influence the direction of (de)regulation in the introduction of ridesharing and AD. On the one hand, the taxi industry association has been against private-use onerous transportation/carpooling and has succeeded in not introducing it, except in certain remote areas. On the other hand, some large taxi companies welcome the introduction of AD, which would help the situation of aging drivers and labor shortages in the industry, and are actively involved in pilot AD projects. In governmental science and technology projects, the close relations between the state, manufacturers, and a few powerful universities have supported the promotion of national projects. Taking advantage of these close relations, the government promotes "all-Japan-efforts" by industry–academia–government collaboration in governmental projects such as SIP-adus (Cross-Ministerial Strategic Innovation Promotion Program, Automated Driving for Universal Service), especially in new innovation frontiers, or common infrastructure technologies that do not involve the private sector's area of competition [24].

2.1.2   State-Citizen Relations in Democratic Countries

Both Japan and Germany are described as democratic countries, if not normatively in Japan, at least operationally so. However, the two countries' democracies differ in specific aspects, and thus the manner in which government power is exercised in AD-related policy processes may vary. According to Merkel [31], five interdependent and independent elements comprise democracy: electoral regime, political rights, civil rights, division of powers (horizontal accountability), and effective power to govern. First, in Japan, division of powers and horizontal accountability function relatively weakly. The judiciary has not actively engaged in politics. Japanese courts have been very conservative and inactive in judicial reviews. Except for a very few cases, the Supreme Court has seldom held government actions unconstitutional, and has maintained its position to avoid constitutional challenges by readily accepting the arguments of the government, or fully respecting the decisions of the legislature [30]. In addition to the historical conservatism of the judiciary, deliberations in legislative bodies, to the extent that they are open to the public, are not very active, and the balance to reciprocally check each other has deteriorated in recent years [37]. All of these factors have opened a way to top-down governance within the institution, with the strengthened function of the executive body in recent years. Furthermore, in Japan, it is not always the elected representatives who have executed the power to govern. This means that public officials and industrial players have historically had a powerful role in certain policy domains. While purportedly representing the model of a developmental state or state-led initiatives, behind the scenes such extra-constitutional actors—who are not directly subject to democratic accountability—have been given leeway to exert their influence. How they are involved in policy processes and the change in power relationships among those actors are explained in detail in the case of AD policies in Sect. 3.

In Japan, even after forming a democratic system, the norms and values that citizens should practice on a daily basis in democratic thinking and civil behavior have not flourished [29]. Consequently, both inside the policy process and through the public sphere, the role of civil society with normative and participatory potential, which can act against authoritarian inclinations, has been comparatively marginal. In fact, only a minority of non-profit organizations engage in advocacy against the government [42]. Street protests rarely mobilize and, in most cases, have no power to change the direction of policy. In addition, Japan's low ranking in the Press Freedom Index for a democratic country indicates a problematic situation regarding the freedom of journalists and news organizations, while some prime ministers in the past have made a long-term government possible partly by keeping the mass media on their side, or by putting pressure on their appointments [57]. Furthermore, against the backdrop that the national government has positioned public transportation as a for-profit business, and that the history of collaboration between citizens and governments in regional transportation planning has been shallow, citizens have had limited channels for involvement in policymaking. In 2010, the Democratic Party of Japan and the Social Democratic Party, then in power, attempted to pass a basic transportation bill guaranteeing the "right to mobility/transport" as a government bill, but due to fierce resistance from different parties, automakers, and public transport service operators, the right was never included in the bill. The government was concerned that it would be accused of inaction as there was insufficient financial support for service operators to guarantee such rights. Local bus companies and regional railroad companies opposed the original bill despite the expected benefits of government subsidies, because their freedom to withdraw from unprofitable routes would be threatened, and the mandatory maintenance of such routes would not contribute to their business stability [23]. Owing to the absence of such rights, it is highly unlikely that citizens will make the lack of public transportation the subject of administrative court cases.

Meanwhile, the population's loyalty to a single political party, the Liberal Democratic Party (LDP), over decades has ensured unwavering control over policymaking with few interruptions, except for a very brief handover of power to other parties [39]. There has been an agreement between the state, business, and the public in Japan that state-led capitalism and technological development are good for the collective interest of national growth, and therefore the developmental state model has been widely accepted. In this regard, the government has made concerted efforts to cultivate loyalty from citizens, and the narrative of the path of shared growth has enjoyed a considerable degree of popular legitimacy [17]. This capitalist regime of pursuing cycles of shared efforts for growth and shared distribution which subordinate social policy to the overriding policy objective of economic growth is called productivist welfare capitalism [16]. This does not mean, however, that social objectives were disregarded; on the contrary, social objectives such as strong firms, technological self-sufficiency, and a cohesive community have been the top priorities. Grounded on the communitarian norms, governmental leaders have pursued policies that represented a wider consensus about what was important [1]. Although support for the developmental regime seems to have decreased since the economic slowdown of the 1990s and worsening technological self-sufficiency, citizens' criticisms may be directed at governmental measures, but not at the governmental role in paternalistic leadership [17]. In Japan, citizen dialogue aims to foster citizens' (or probably rather future consumers'/users') understanding and acceptance of AD, and the government may play a director-like role in organizing such communication activities.

In the German tradition, it is generally agreed that the division of the state powers of legislative, executive, and judiciary bodies, and the horizontal accountability among the powers are an important part of the rule of law and democracy [31]. Following the Nazi era, one of the elementary goals was to design the Basic Law of the Federal Republic of Germany in such a way as to avoid an excessive concentration of power. The responsiveness and responsibility of the government are considered to be secured through mutual checks and balances, and therefore, the exercise of executive power appears to be more balanced and limited when compared to Japan. Although German legislation does not establish the right to transportation (for example as is the case in France), the competent authorities have legal responsibilities for the population to be adequately served by local public transportation services, in line with the principles of climate protection and sustainability. German citizens, especially those with reduced mobility or sensory impairments, are given several formal and informal opportunities to participate in the formulation process of local transport plans (see chapter "Social Acceptance of CAD in Japan and Germany: Conceptual Issues and Empirical Insights" of this book regarding the high level of support among the German public for involvement in the planning of autonomous vehicles). In addition to formal participation in transportation and land-use planning, citizens' input has also been reflected in discourses and framing of technological innovation through social movements since the 1970s [4]. Policymakers are sensitized by public reactions, and social dialogue is often considered essential to the success and social acceptance of emerging technologies and new planning. In these cases, dialogue is viewed as a process of negotiation in which benefits and risks are carefully deliberated, rather than as mere communication to build trust among stakeholders and justify decisions. In principle, citizens are expected to be *citoyens* whose civic duty is to participate responsibly in democratic decisions and public life, including the assessment and governance of technology. In Germany, the role of government is imagined as a mediator that facilitates social negotiations by enabling exchanges among citizens, science, and industry. This process is believed to enhance the democratic legitimacy of policy decisions in emerging technologies such as AD. This institutionalized practice of consensus-seeking for collective choices is a result of institutional traditions established after World War II, in which creating and maintaining social equilibrium has been deeply interwoven in political and economic governance [4].

2.2 Polity: Institutional Settings in Political Arrangements

State actions are influenced by different political arrangements in a country. That is, whether it is a single authority in unitary institutional structures, or a federal or regional institutional structure distributed among multiple authorities; whether it is a statist or more corporatist policymaking process; whether it is a system of majority representation or proportional representation; and how the public participates in politics [45]. While political decision-making arenas for the transportation system have been rather centralized in Japan, policymakers in Germany are required to formulate relevant policies in multi-level governance—at the European, national, state, and local levels. The European Union (EU) actors, such as the EU Commission, the European Court of Justice, and the European Parliament, the federal government, and local authorities influence one another on relevant regulations and transportation system developments. To facilitate vertical coordination at different levels, the federal government may need to act as a policy moderator rather than a policymaker, and it has a role as an advocate for local communities vis-à-vis the EU Commission [2]. Moreover, in a coalition government, negotiations between parties with different interests are inevitable, especially on topics that straddle different policy areas, such as AD. Additionally, although the polity in both countries is democracy, the two countries' democracies differ in specific aspects, as explained in Sect. 2.1.2. The state-citizen relations and the nature of civic participation in each country differ due to the different institutional traditions. This political institutional context provides an explanation for the coordination mechanisms among political actors, and whether the state can exercise its power to impose policies relatively easily [45].

2.3 Policies: Interpretation of AD in Different Policy Areas

The topic of AD has been discussed in different policy areas, such as public services, industry, and science and technology. How AD is conceptualized and interpreted in different policy areas is influenced by how the government conceives of public transport and science and technology in relation to the state and its population. Accordingly, governments may take different substantive political actions and strategies. For example, in Japan, the public transport sector, including railroads, buses, and taxis, was a main target of privatization and deregulation from the 1980s to the early 2000s to improve profitability and competitiveness, which has resulted in the majority of transportation services being provided by private companies. 1987 saw the privatization of Japan National Railways (JNR) and its division into JR companies. Then, in 1996, the Ministry of Transportation (now the Ministry of Land, Infrastructure, Transport and Tourism, MLIT) decided to loosen the supply–demand adjustment regulations for all modes of public transportation. In accordance with this decision, bus deregulation measures were implemented in 2002 to lift the permit system for entry and exit from the business, resulting in the withdrawal of bus operations in rural and suburban areas [47]. Public transportation has been considered as a for-profit business for a long time, and the role of national and prefectural governments has been limited mainly to supervising and coordinating operators [41, 48]. However, the decline in public transportation in recent years has been interpreted as a decrease in the vitality and productivity of local residents in their economic and social activities, and the government has once again taken up public transportation as a policy concern. Thus, the transportation sector has been discussed in Japan in the context of the profitability of the operators, and the vitality and productivity of residents through mobility. Responding to the situation of declining public transportation in rural and suburban areas, the "Basic Act on Transportation Policy" was enforced in 2013, and the "Act on Revitalization and Rehabilitation of Local Public Transportation" was amended in 2014. These clarified the responsibilities of the national government, prefectural governments, local governments, operators, and citizens in public transportation policy, and as a result local governments can now take more proactive roles in public transportation policy by establishing a "regional public transportation network formation plan." Yet, this is still a new trend, and many municipalities have difficulties in responding immediately to the given responsibility in terms of their institutional structure, personnel, and budgets [48].

In Germany, public transport services have been consistently interpreted as elementary services of public interest, and the state has a duty to ensure mobility in accordance with the welfare state principle (Article 20 (1) of the Basic Law for the Federal Republic of Germany). In 1938, the concept of Daseinsvorsorge (public service) was proposed by Ernst Forsthoff, a German scholar of constitutional law, with the intention of clarifying the relationship between the individual and the service-providing state. While he noted that it is impossible to limit the scope of Daseinsvorsorge in quantitative and qualitative terms, the provision of public transportation was named as one such service. This idea continued in administrative practice in the post-war period. In 1993, the Law on the Regionalization of Public Transport was enacted, which specified the responsibility of authorities to secure an adequate level of public transport services for the population in their territories in accordance with the provision of services of general interest (§ 1 Regionalisierungsgesetz—RegG). This federal law decentralized the responsibility for planning, organization, and financing of local public transportation to regional governments (§ 3 RegG). Each regional government has the further authority to designate administrative bodies to carry out the administrative practice of urban transportation, and the designated local agencies are responsible for ensuring that residents are adequately served by local public transportation services in line with the principles of climate protection and sustainability (§ 8 Personenbeförderungsgesetz—PBefG). Thus, the concept of Daseinsvorsorge, hitherto legally undefined, requires competent regional and local authorities to guarantee the provision of adequate short-distance passenger transportation services that meet the ordinary mobility needs of individual population groups, even if the provision of these services is not undertaken by the governments themselves or by public enterprises. Therefore, the provision of services is not categorized as a private activity, as it is in Japan, and the German public transportation system has only partially opened up to direct competition amongst companies. The competition is highly regulated within the framework of a heavily-substituted service sector [2]. The role and responsibilities of public entities in public transportation appear to be significant, in that securing provision of public passenger transport services remains, in principle, the responsibility of the state.

In accordance with the principles of climate protection and sustainability, many policymakers in Germany tend to consider new mobility services, such as AD taxis and shuttles, as complements to, not replacements for, the public transportation network. In this context, it should be noted that a new legal regulation related to AD was approved in Germany in 2021. Under this law, Level 4 vehicles (for levels of AD see the introduction) are principally allowed to operate in mixed traffic, on public roads, in predetermined areas in Germany. The AD vehicles are allowed to operate without a driver on board. However, a technical supervisor who can deactivate or enable driving maneuvers of the vehicle from outside is necessary. It is possible for one person to simultaneously supervise several vehicles. A directive specifying the implementation of the new law was approved in May 2022. It is assumed that the new law particularly supports the integration of Level 4 vehicles into public transport services. Public transport operators are considered particularly suitable to meet requirements such as the provision of a supervisor. Many experts expect this to be the first step towards significantly more sustainable and less carbon-intensive mobility based on a new generation of public transport services [5]. Thus, the differences in public transportation policies in Germany and Japan stem to a large extent from the different ways of conceptualizing transport or mobility in each country, which in turn may result in different actor constellations and different approaches to the social implementation of AD. In Japan, privatization and regulatory measures have influenced the public transportation sector, with the aim of enhancing capitalism through state intervention in private activities, which is a typical approach by SMEs. However, in addition to the shallow history of local transportation planning, privatization has created a situation where it is rather difficult for the state to take strong initiatives in comprehensive mobility system development, including the public transportation network. Although the state seems to continue to hold the strong idea of improving economic efficiency in the mobility sector through state intervention, discussions on how AD should be integrated into the public transportation network cannot proceed without communication and cooperation with private companies and for-profit public transport operators. In Germany, the state has different motivations to engage in public transportation than is found in SMEs, in that it interprets public transportation as a fundamental public interest in the welfare state. However, policymakers who have been utilizing different policy instruments, such as regulations, subsidies, and transportation planning, may be able to provide direction more proactively in discussions on how AD can be integrated into the public transportation network. Furthermore, different development pathways of AD may have different consequences, depending on whether, for example, public transport companies in Germany would play a leading role in the integration of AD vehicles and services.

Moreover, the interpretation of the role of science and technology in society is not the same in both countries. In Japan, the liberalization policy has resulted in a limited, indirect government role for public transportation, while its role in the development of science and technology has been consistently emphasized. In the past, the national government has taken the initiative in the import and development of basic and applied science, recognizing the need to catch up with industrialized and advanced nations with the greatest speed [11]. Scientific and technological independence, or technological superiority, has been considered to improve national security, and this technonationalism has been a key concept in Japan. In such technonational regimes, the government assumes responsibility for guiding industries and markets. It is not that Western countries have never pursued national security through technological leadership, but Japan, like other East Asian countries, may represent an extreme case whereby a national technological development vision is supported by government leaders as well as domestic enterprises [22]. In the 1980 edition of the White Paper on Science and Technology formulated by the former Japanese Science and Technology Agency, a vision of "*Kagaku Gijutsu Rikkoku*" (nation-building by science and technology) is mentioned as a national goal, and became a long-lasting slogan of the Japanese government. Furthermore, recent government innovation policies are based on the idea that the ability to generate technological innovation is essential to sustain economic growth [8]. Against this backdrop, AD has often been discussed in science, technology and innovation policies in Japan over the past decade. In more recent years, AD began to be discussed more often in the context of public transportation, as the social issue of public transportation withdrawal in remote areas has emerged, and practical application of AD is expected in the service sector sooner than in the private car sector. Yet, policies on mobility tend to be subordinate to the national vision of a smart and digital society based on new science and technology; AD is first and foremost contextualized in Japan in relation to new technology and innovation that should improve national security, promote economic growth, and solve social issues. In this way, state actions on AD are influenced by government expectations of science, technology and innovation, and the expected role of the government in realizing those expectations.

Driven by the expectations that science and technological superiority and the ability to generate technological innovations will improve national security and sustain future economic growth, the Japanese government has defined its own role in realizing these expectations by demonstrating leadership. However, the policy has also changed in response to the perceived needs of the economy at a given point in time [38]. During the "catch-up" period (1950s to the early 1970s), which brought about the economic miracle, the state played the most direct role in economic coordination. It was argued that national technological upgrading could benefit from policies that emphasized cost–benefit considerations and performance criteria, and thus could moderate the distorting effects of policy interventions [8]. Later, in the 1980s, the weight of R&D shifted from the improvement of imported technologies to domestic invention, and companies were expected to innovate within their own organizations, using their own capacities. Since 1990, during the recessionary period after the burst of the bubble economy, the state has recognized the importance of a growth strategy through science-based innovation, in which the research outcome should be applied and developed for industrialization [38]. It then increased government spending on R&D to the level of major Western countries. However, increasing budget deficits, coupled with the rather unsatisfactory outcome of the attempt to introduce a market-based framework of innovation systems by limiting the government's role, led the government to adopt a more top-down style of priority-setting and planning in R&D and social implementation. From this perspective, science-based innovation, including information technology, has been prioritized [24, 25]. In this regard, automation technologies are recognized as an important technological area by the government, in terms of the interaction between scientific activities and industrial innovation. Thus, Japan maintains its SME character in science, technology and innovation policy, except for the brief attempt at a market-based framework for an innovation system.

In Germany, the expectations of science and technology are twofold: economic welfare and environmental improvement (or sustainability in a broader sense). Scientific and technological innovations are expected to advance economic progress, while diminishing the unintended side effects of modernity [4]. Although both Germany and Japan have similar expectations of economic advancement through the promotion of science and technology, there is a nuanced difference in that in Japan, the expectations for economic growth through technological superiority are interrelated with national (economic) security. In Germany, the expectations of environmental improvement by innovative technologies seem to be largely shared by both policymakers and citizens (see chapter "Social Acceptance of CAD in Japan and Germany: Conceptual Issues and Empirical Insights" regarding German citizens' expectation of environmental improvement through AD), as a result of the success of ecological movements since the 1980s in incorporating environmental problems into the social and political discourse of technological innovation. The coalition agreement of the new German government signed in November 2021 states: "Mobility should be made sustainable, efficient, barrier-free, intelligent and affordable for all." The decarbonization of the mobility sector is mentioned as an overall objective. Accordingly, the national strategy on AD as a technological innovation assigns a relatively high degree of importance to economic and environmental issues. The role of the government includes deliberation on the potential positive and negative consequences of emerging technologies, which are often assessed and governed in accordance with the precautionary principle [4]. In this way, climate protection is a concern in both topics of public transportation and scientific and technological innovations, and AD is often closely interconnected with environmental policy in Germany.

Actors' relations, influenced by the roles expected of them by others (as discussed in Sect. 2.1), can be seen to differ across policy areas. In Japan, the government believes in its responsibility to guide industries and markets in a technonational regime, and industry and the public also expect the government to exercise leadership in economic and science and technology policies by setting an overall goal. In Germany, in transportation and science and technology policies, the government serves rather as a mediator or moderator in social and political negotiations with citizens and other political actors at different levels. The responsibility for ensuring public transportation services is placed on the government, and German citizens therefore expect the government to play a more prominent role in public transportation policy than Japanese citizens probably do. At the same time, in Germany, citizens are expected to participate responsibly in democratic decisions on public life, where both technology and mobility are relevant issues. Such expectations of each other's roles in these different policy areas influence how these actors interact and whether the resulting political decisions will be accepted by them.

In this Section, in order to understand the differences between state actions and governance styles in AD in Japan and Germany, we presented different aspects in politics, polity, and policies: actor relations influenced by their expected roles; centralized or multi-level, and contrasting democratic political arrangements; and contextualization and conceptualization of AD related to different interpretations of technological innovation and public transport. Section 3 examines the policymaking arena and the central actors in Japan, which have changed significantly since the last century, with examples of AD-related policies. Based on a literature review of past and recent Japanese policymaking, policy documents, strategy papers, meeting minutes, and government websites were investigated, with a particular focus on the period of former Prime Minister Shinzo Abe (2012–2020).

3 Policy Processes, Power Relations and the Recent Changes in Japan

3.1 Traditional Policy Process

The developmental state, or the Japan Inc. model, in the post-war period was featured by the mutually dependent, reciprocal relationship among interest groups, politicians and bureaucrats described as *"Iron Triangles."* To many outsiders, this appeared like collusion, because new proposals were submitted for approval to the Cabinet and to Parliament only after agreement was reached among them [8]. The balanced equilibrium among institutions, policies, and socioeconomic blocs has provided a "positive cycle of reinforcing dominance" [39]. In the institutionalized mode of cooperation on technological upgrading, the Ministry of International Trade and Industry (MITI, now the Ministry of Economy, Trade and Industry, METI), together with the Ministry of Finance, served as the main conduit for economic governance throughout the second half of the 20th century [17]. MITI played a significant role in shaping future directions by identifying future prospects for social and economic needs, combined with approaches such as technology foresight and close cooperation with related industrial policies. During the period of catching up with the West, until the early 1980s the Japanese government and industry mostly shared the same expectations, preferences and possibilities with respect to their futures, based on a mutual goal of wealth creation through technological development. Wealth was commonly defined in terms of capital accumulation and creation of value-added goods and services. In line with this consensus, industrial and academic contributors to vision formulation were in return given the opportunity to participate in national R&D programs to realize the vision [56]. The "eye-catching" and "reliable" future visions of the government, according to Shinji Fukukawa, former vice-minister of MITI, had served as the guideline for the private sector and gave dynamism in the business direction [10]. This outstanding presence of MITI in post-war industrial policy, which was considered to have contributed to the miraculous economic growth in Japan, was featured in Chalmers Johnson's 1982 work, *MITI and the Japanese Miracle* [26].

The ministries utilized a council that was attached to each administrative agency of the state, whose bureaucratic servants virtually controlled policy through the councils, by holding important secretariat functions such as selecting council members and setting the agenda [37]. Meanwhile, the LDP exercised political oversight largely through the functionally-specific committees of its Policy Affairs Research Council (PARC) in close cooperation with bureaucratic agencies. The problem with this vertically separated council system was that it did not provide an opportunity for an independent cabinet to initiate legislation, or a policy arena for the comprehensive horizontal coordination of ministry boundaries and interrelated topics, which had not been a problem when there was a strong consensus regarding catch-up [37, 39]. This policymaking environment has, however, experienced gradual but radical change since the turn of the 21st century, following the collapse of the bubble and a series of scandals involving élite bureaucrats, as well as incessant waves of globalization [26].

3.2 Power-Shifting to the Prime Minister's Office

Problem-awareness that the government officials who should have been assisting in policy planning and execution led the politics, and the government was too dependent on them—the LDP's "headless" government (Mishima [32], p. 105)—was widely shared when entering an era of low growth and persistent budget deficits in the 1990s. However, with the purpose of increasing the leadership of politicians, the power was shifted to a core executive centered on the Prime Minister (not to Japan's bicameral parliament, the National Diet), an approach known as *"Kantei Shudō"* (the Prime Minister's Office's leadership), and thereby the era when it was ridiculed as *"Kanryō Shudō"* (bureaucratic leadership) in the last century came to an end [3, 33, 37]. The state's central organization went through several steps to strengthen its political leadership. The first step was electoral reforms in the early 1990s, the second was the Hashimoto Cabinet's administrative reforms in the late 1990s, and the third was reforms of the civil service system, which began in earnest at the beginning of the 21st century, and eventually established centralized control of senior civil servants by the Prime Minister's Office [37], Preface). In brief, the reforms of the election system and the political funding system in 1994 resulted in a weakening of *"Habatsu"* (political factions in the party) of the LDP, as well as *"Zoku gi'in"* (parliamentarians who specialize in a political area, and therefore have strong connections with the private sector and departments in ministries of that area), and concentrating the power and money in hands of party leaders [32]. As a consequence, bargaining and competition among the factions and *"Zoku"* in LPD, which used to activate discussions in the PARC (called *"Seichōkai"* in LDP) in order to coordinate different interests within the party, also receded [37].

Through the administrative reforms of the Hashimoto Cabinet in the late 1990s, which aimed to eliminate the negative effects of vertically divided administration and strengthen government functions, the new structure of government ministries and agencies started in 2001. In these reforms, the number of councils in ministries, where public officials previously exercised their power, decreased. Instead, the Cabinet Office was given responsibility for the general coordination of matters for which it was difficult to specify the dedicated ministry or agency, and the Cabinet Secretariat was confirmed to be responsible for the overall coordination from the perspective of directly assisting the Prime Minister as the highest and final coordination body under the Cabinet [46]. This led to the erosion of many of the previously close ties between ministries and interest groups [39]. At this time, the Prime Minister was also given explicit authority to request an agenda for Cabinet meetings and to initiate legislation, and some posts for political appointments of career bureaucrats were newly established, such as the Special Advisor to the Prime Minister, the Assistant Chief Cabinet Secretary, and the Cabinet Public Relations Secretary [37, 46]. These reforms created a foundation for a form of governance that enabled the Prime Minister and other core executives in the Cabinet Secretariat, as well as some ministers, to take leadership. Then, the policymaking mechanism through policy conferences was institutionalized through the political leadership of Prime Ministers such as Junichiro Koizumi (2001–2006) and Shinzo Abe (2006–2007, 2012–2020), and by then Democratic Party of Japan when it was in power (2009–2012).

Currently, political strategies on important topics are developed mainly in "*Seisaku Kaigi*" (policy conferences). Although there is no official definition of policy conferences, mainly two different types of exist: the five councils on important policies that were established in the Cabinet Office as a result of the Central Government Reform in 2001, and those designated as policy conferences among various bodies operated directly by the Prime Minister's Office or the Cabinet Secretariat [37]. Unlike those in the first group, the establishment of the second does not require a decree, and flexible setting-up by the Cabinet is possible. As a consequence of this flexibility, the number of policy conferences has increased from only 39 in the Mori Cabinet (2000–2001), to 168 in the second Abe Cabinet (2012–2014), which includes 114 newly established conferences during his time (Nonaka and Aoki [37], chapter "Setting the Scene for Automated Mobility: A Comparative Introduction to the Mobility Systems in Germany and Japan", Sect. 2).

The efforts by the Japanese government to reduce bureaucrats' dominance in the policymaking process have made significant changes in their relations. Especially, the establishment of the Cabinet Bureau of Personnel Affairs in 2014 by then Prime Minister Shinzo Abe seems to have had an impact on power relations. At this time, it was decided that the Chief Cabinet Secretary had the right to create a list of candidates for approximately 600 senior positions in the civil service, and ministers were required to consult with the Prime Minister and the Chief Cabinet Secretary prior to appointments, dismissals and demotions [37]. Previously, each ministry and agency had the right to prepare a list of candidates for approximately 200 positions of chief of bureaux. Through this strengthened authority over appointments and dismissals, the Prime Minister and the key members of the Cabinet Secretariat gained more control over governmental officials. Furthermore, following the condition that the number of career bureaucrats working in the Cabinet Secretariat can be flexibly determined by a Cabinet order, the staff number has increased significantly from 1,054 in 2001 to 2,929 in 2015. In 2015, about one-third were whole-time positions in the Cabinet Secretariat, another one-third were concurrent staff members who kept their position in the ministry or agency but were stationed in the Cabinet Secretariat, and the remaining third were concurrent employees who were stationed mainly in their ministry or agency [46]. Today, the Cabinet can easily gather preferred personnel for policy conferences, depending on the topics of interest.

In most cases, it is still bureaucrats who write policy proposals as they did when it was called "*Kanryō Shudō*"; however, they now work in the secretariats of policy conferences directly run by the Cabinet Secretariat or in the Cabinet Office. They are under regular supervision by the Prime Minister, Chief Cabinet Secretary, and responsible ministers, making it impossible for them to avoid oversight [32]. Ultimately, "the discussions in conferences are more like a ritual," since the overall policy direction is more or less previously decided by the team in the Prime Minister's Office (Tanaka [50], p. 75). There is no change to the main idea in policy proposals prepared by selected bureaucrats working in the secretariat of policy conferences, and the meetings are rather a place to justify the proposals submitted by invited experts [27]. Indeed, conversations in published minutes of meetings of the Road Traffic Working Group under a policy conference called IT Strategy Headquarters, which is supposed to formulate Public–Private ITS Initiative/Roadmaps, suggest that expert members from academia and industry simply comment on drafts already prepared by the secretariat.

Depending on the policy focus of each Prime Minister, some bureaucratic officials from ministries would be assigned to higher or closer positions to the Prime Minister, and therefore they have opportunities to influence policy direction. During his second tenure of office, bureaucratic officials from METI were in favor of Prime Minister Shinzo Abe; for example, Takaya Imai, a civil servant from the ministry, held a key position of Executive Secretary to the Prime Minister, as well as Special Advisor to the Prime Minister [35]. On the topic of AD, bureaucratic officials from departments that have conventionally been involved in policy areas of road traffic, road transport, and automobiles seem to still play significant roles by being summoned to policy conference meetings. For example, bureaucratic officials from the following divisions or positions often attend meetings of AD-related subordinate bodies under policy conferences: Automobile Division, Manufacturing Industries Bureau, METI; Road Traffic Control Division, Road Bureau, MLIT; Engineering and Environmental Policy Division, Road Transport Bureau, MLIT; Land Mobile Communications Division, Telecommunications Bureau, Ministry of Internal Affairs and Communications; Traffic Planning Division, Traffic Bureau, National Police Agency (NPA); and the Counselor of the Commissioner-General's Secretariat of the National Police Agency. This wide range of departments in charge shows the vestiges of a vertically-divided administration, and AD-related topics involve the territories of different ministers and agencies that were previously siloed. Bureaucratic officials who are called to policy conferences will try to maximize the interests of their own ministry or department, while conjecturing the Cabinet's desire in the overall direction of the policy topics of the conferences.

Although the influence of bureaucracy is somewhat preserved, it is now under the leadership of the core executive, centered on the Prime Minister. The core executive is a close circle of key central actors, composing an asymmetric position of dominance over other actors in the policymaking arena [33]. In terms of AD-related policy, core executive such as the Prime Minister, relevant Ministers of State, other politically appointed officials, such as the Chief Cabinet Secretary, appointed bureaucratic officials from METI, MLIT, and NPA, seem to play a significant role. In topics such as AD, in which different authority departments and industrial players have a strong interest, one wonders how the circle of the core executive can have strong leadership and proceed in a fixed-game manner in the direction and discussions of policy conferences. There is likely to be an undisclosed process for consensus building, known as "*Nemawashi*," outside the officially-recorded policymaking process. For example, according to Takenaka [49], Minister of State for Economic and Fiscal Policy in the Koizumi administration (2001–2006), secret strategy meetings of the so-called CPU (Communication and Policy Unit) were held every Sunday by a close circle of the core executive, such as the Minister of State for Economic and Fiscal Policy, the Chief Cabinet Secretary and Deputy Chief Cabinet Secretary, and a key member from a business field in the Council on Economic and Fiscal Policy. This CPU seemed to have continued during the second Abe administration. In addition, meetings were organized every day after lunch by the Abe Cabinet, attended by a small circle including the Prime Minister, the Chief Cabinet Secretary, three Deputy Chief Cabinet Secretaries, and the Executive Secretary to the Prime Minister, Takaya Imai from METI (Taniguchi [52], chapter 3, Sect. 1). This suggests very close communication among the important figures of the core executive. Furthermore, then Prime Minister Shinzo Abe had frequent visits to his office by bureaucratic officials and the private sector[2] (for more unrecorded meetings, see Machidori [28]). Thus, policy drafts may be written through meticulous preparation by the close members of the Prime Minister and bureaucratic officials who can surmise their superiors' wishes based on discussions between the core executive and industrial players outside the policy conferences. Policy documents are the result of such opaque, often undisclosed coordination processes to reflect the interests of stakeholders, and the resulting economic and industrial policies are promoted under the name of political leadership with the support of the industry.

---

[2] The comings and goings of visitors to the Prime Minister's official residence are recorded by news agencies throughout the day, and the data are distributed to daily newspapers. According to the Asahi Shimbun, then Prime Minister Abe had visits 157 times by public officials and 67 times from the private sector during the period of the first month in his second Cabinet. The summed number of visits is the highest compared to the equivalent month of the previous nine prime ministers since 2000. The trend of more visits by public officials and private sector representatives and fewer visits by politicians continues in the following Suga Cabinet. (https://www.asahi.com/special/shu sho-1month/, accessed on August 23. 2021).

When the structure enabling the Cabinet to execute leadership in the policy direction was set up, the main (officially-recorded) policymaking arena shifted from ministry-based deliberation councils to policy conferences under the Cabinet or the Cabinet Office. As a consequence, this allows the Cabinet to pursue cross-cutting and innovative policy goals, unlike the previous "*Kanryō Shudō*" policymaking that constituted more or less minor adjustments within policy areas where the ministries had vested interests and territories, and therefore did not necessarily need clear national vision across different ministries [37]. The recent policy process seems to be facilitated by selected bureaucratic officials from specific ministries and agencies working for policy conferences who understand the Cabinet's intentions, as well as invited industrial players who are like-minded about economic policy. The cooperative top-down policymaking process is, however, far from transparent, with deliberations by active parliamentary politics reflecting the public will more comprehensively. How the National Diet is involved in the topic of AD is further discussed in Sect. 3.5 regarding regulatory processes.

3.3   AD Policy Process Through Policy Conferences Under Prime Minister Shinzo Abe

The following policy conferences have published policy documents or strategy papers relevant to AD: the Headquarters for Japan's Economic Revitalization (abolished in 2020); the Council on Overcoming Population Decline and Vitalizing Local Economy; the Integrated Innovation Strategy Promotion Council; the IT Strategy Headquarters; and the Council for Science, Technology and Innovation (see Appendix in Table 1). Among the policy conferences, the first three were established by Prime Minister Shinzo Abe in his second tenure (2012–2020). One focus was to boost economic growth, which was largely supported in his election in 2012. After taking office he promoted tri-partite economic strategies known as "Abenomics" [3]. His strong political interest in the economy appeared, for instance, in that setting up the Headquarters for Japan's Economic Revitalization under the Cabinet[3] was decided in a Cabinet meeting on the day of his inauguration, despite a similar policy conference (the Council on Economic and Fiscal Policy), existing in the Cabinet Office since 2001. In addition, the Cabinet often insisted on strengthening the function of policy conferences as a command/control tower ("*Shireitō*"). As a consequence, the Japanese name of the IT Strategy Headquarters was changed to the IT "*Sōgō*" (comprehensive) Strategy Headquarters,[4] which confirmed that it should work to control the overall IT strategies of the government (IT Strategy Headquarters [19], p. 2). The Council for Science and Technology, whose predecessor had existed since 1959, also changed to the Council for Science, Technology and Innovation in 2014,

---

[3] The policy conference was abolished by the following Prime Minister, Yoshihide Suga, on October 16, 2020.

[4] The English name of the conference was left unchanged as the IT Strategy Headquarters.

which indicates the efforts to increase the controlling power over innovation by the Cabinet. Furthermore, the Integrated Innovation Strategy Promotion Council was established in 2018 directly under the Cabinet, despite the existing similar conference in the Cabinet Office; this was justified to improve the coordination of several *"Shireitō"* conferences related to innovation, including the Council for Science, Technology and Innovation, and the IT Strategy Headquarters (Government of Japan [13], p. 5).

Even though different AD-relevant policy conferences (see Appendix in Table 1) have their own topics to focus on, the Cabinet's interest in economic growth overarches them. The IT strategy was counted as a pillar of the growth strategy, which was one of Abe's tri-partite economic strategies (IT Strategy Headquarters [18], p. 2, 3). In addition, the second stage of "Abenomics," announced in 2015, tried to bring more attention to demographic changes from an economic perspective. The dwindling birth rate and aging population were reaffirmed as significant issues, in terms of the decreasing size of the working population, which is a decisive factor for economic growth, besides other factors such as capital stock and total factor productivity. This viewpoint linking demographic change to economic issues appears in the name of a conference: the Council on Overcoming Population Decline and Vitalizing Local Economy. Innovation was also recognized as a key to the economy, just as science and technology have always been identified as playing a vital role in industrial development and the economy. Based on the economic growth model from neoclassical economics, the Integrated Innovation Strategy 2020 also argues that the promotion of innovation is significant to increase total factor productivity, which contributes to economic growth, even under the condition that the increase of the (working) population in the future cannot be expected (Government of Japan [14], p. 7).

In those policy conferences with strong interests in the economy, technological development and system introduction of AD have been promoted with expectations to contribute to economic growth. At the end of 2017, the New Economic Policy Package was released, which was formulated in the Headquarters for Japan's Economic Revitalization and approved in a Cabinet meeting. It set the next three years until 2020 to work on *"Seisansei Kakumei"* (literally meaning productivity revolution, translated to "Supply System Innovation" in the English versions of the document) to raise the productivity of the entire Japanese economy. AD was mentioned in the section of "Societal implementation of the Fourth Industrial Revolution and the system reforms in the areas experiencing sluggish productivity" (Government of Japan [12], pp. 1–1, pp. 3–6). This New Economic Policy Package also decided to start the second Cross-Ministerial Strategic Innovation Promotion Program (SIP) ahead of the original schedule. Based on this push, the 12 issues, including AD continued from the first SIP, were formalized at the Council for Science, Technology and Innovation (IT Strategy Headquarters [21], p. 1). This sequence suggests that the policy conference of economic revitalization was in a strong position compared to other conferences, and the productivity improvement by AD seems to be a core promise shared by different policy conferences.

3.4 Continued Contribution by Industrial Players

Industrial players and interest groups had been entangled in the Japanese policy network as a part of the *"Iron Triangles"*, which they composed with LDP's *"Zoku gi'in"* (parliamentarians) and bureaucrats (relevant authorities) [51]. But they did not lose their long-standing institutional ties to the government, even after many of them had grown sufficiently to compete in the international market, and the custom of large companies participating in national projects continued [17]. As the nation's economic power stagnated, industrial players recognized the absence of political leadership as a "terrible problem," and retained the expectation that the state should provide direction and leadership for the economy (Witt and Redding [59], p. 873). While the power of *"Zoku gi'in"* and bureaucrats were fractured to some extent through the reforms with the emergence of the Cabinet described in Sect. 3.2, industrial players seem to keep a strong channel to execute influence over policymaking.

In some policy conferences, experts including industrial players who are not members of the National Diet, have been appointed to key positions. For example, in the IT Strategy Headquarters and the Council for Science, Technology and Innovation, experts from business fields are involved as members in the same position as some ministers. In subordinate meeting bodies of policy conferences, original equipment manufacturers (OEMs), automobile-related industrial associations and internet and telecommunication companies seem to dominate, while few conventional mobility service providers and consumer organizations participate (see Appendix in Table 1). The members from industrial fields are well-represented not only in policymaking, but also in the social implementation of AD by the government. For example, as of October 2018, the executive chairperson of DeNA Co., Ltd. was a member of a policy conference in which the Growth Strategies were drafted, and the mobile internet company was also counted as a member of the Road Traffic WG under the IT Strategy Headquarters to develop the Public–Private ITS Initiative/ Roadmaps. The company's name also appeared regularly in the practical experiments of AD coordinated by the government. Similarly, Toyota Motor Corporation has been a member of the Council for Science, Technology and Innovation as well as some meeting groups under the IT Strategy Headquarters. In addition, the head of the Advanced R&D and Engineering Company in Toyota is the Program Director of SIP-adus. The selection of experts, especially those in subordinate meeting bodies under the policy conferences, might be rather arbitrary. It does not always require parliamentary approval, because the original purpose of those subordinate meetings is to investigate and research important matters in response to consultations by the Cabinet, and they are not required to make (political) decisions. However, in reality, Cabinet meetings often lack the original function of deliberation, so the reports submitted by the responding policy conferences to the Cabinet are hardly checked, and their decisions are very likely to be the final decision of the government [37]. In this sense, the policy documents that are developed by invited experts who succeed in making their preferred measures fit the overall government goals would easily pass the Cabinet. This is probably facilitated by *"Nemawashi"* through undisclosed, informal visits and meetings between the core executive and industrial players outside the policy conferences. In addition, since similar expert members are summoned in different conferences, the drafted strategies including promotion of AD appear more or less homogeneous in different policy documents.

As mentioned in Sect. 3.3, during the terms in office of Prime Minister Shinzo Abe, policies in different areas, such as IT and science and technology, were promoted from the perspective of their contribution to economic growth and productivity improvement. His strong interest in economic growth, largely supported by the population, contributed to the industry's voice being stronger than ever. According to Kazuo Kyuma, formerly a standing advisor to Mitsubishi Electric Corporation, who served as a full-time member of the Council for Science, Technology and Innovation as of 2014, the national vision "Society 5.0" was developed by members of the conference including himself, Takeshi Uchiyamada (Chairman of Toyota Motor Corporation), the late Hiroaki Nakanishi (then Chairman of Hitachi, Ltd./then Chairman of the Japan Business Federation), as well as the secretariats of the policy conference and some contributors from the business world. The societal future concept was integrated in the 5th Science and Technology Basic Plan (5-year plan 2016–2020), and is often referred to in other policy documents by policy conferences to justify political measures including the promotion of AD. Kazuo Kyuma recalls that the voice of the industrial world was strongly reflected in the 5th Science and Technology Basic Plan, which was different from previous basic plans, in which ideas from academia had more consideration [36]. By supporting the leadership of the Cabinet, the industrial players seem to enjoy positions within or outside policy conferences to develop concrete ideas to contribute to the overall governmental goal of economic growth.

3.5 Rulemaking and (De)regulation

To enable Level 3 AD on public roads in Japan, two Cabinet bills to amend the Road Transport Vehicle Act (RTVA) and the Road Traffic Act (RTA) were submitted to the National Diet in 2019 (for levels of AD see the introduction). MLIT is responsible for the first Act (RTVA), and NPA for the second (RTA). After a push by the Charter for Improvement of Legal System and Environment for Automated Driving Systems published in 2018, drafts of the amended law were prepared by bureaucratic officials of the ministry and the agency respectively, and the submission of the bills to the National Diet was decided by a Cabinet meeting.

The main purpose of the amendment of the RTVA was to set safety standards for the practical use of AD, including setting driving environment conditions by MLIT for each vehicle type, mandatory installation of a device to record driving data, establishment of a permission system for performance changes or modification of programs, and requirements for vehicle manufacturers to provide model-specific specifications to mechanics for inspections and maintenance. The amended RTVA newly defines "automated driving device," and the use of this is stated as "driving" in the RTA. This clarifies that people who use Level 3 automated vehicles will also be regarded as drivers and must fulfil the obligations of a driver, as defined by the RTA. The amendment of the RTA includes obligation to record driving data by drivers, prohibition of operation outside the specified driving environment conditions, and obligations to take over driving authority and duties if the conditions are no longer fulfilled, while allowing drivers to use cellphones or other image display devices under certain conditions.

Japan has a committee-centered system in which the main discussion field in legislation processes open to the public is in committees, rather than plenary sessions [9, 37]. The topic of the amendment of the RTA was brought up on April 11, 2019 in the Cabinet Committee of the House of Councilors. This is one of the 17 Standing Committees that exist to address each policy area in both Upper and Lower Houses respectively; members of the National Diet must belong to one or more of the committees of the House. In the meeting of the Cabinet Committee of the (Upper) House of Councilors, six politicians from different parties each questioned the bill within the time allocated to them, and either the Chairperson of the National Public Safety Commission or bureaucratic officials who attended as reference persons provided answers, without any interruption by other politicians. The Chairperson of the National Public Safety Commission is a minister of state, which is the parent agency of NPA, traditionally in charge of the RTA. The meeting appears to be a question-and-answer session, or fact-checking session, for members of the committee to collect information. There was no time for deeper discussion based on the answers and information provided. Ultimately, the revised draft proposed by an opposition party was rejected and the bill was approved by a majority vote as originally submitted. At the end of the meeting, a supplementary resolution was jointly submitted by parties who expected certain political effects despite there being no legal basis for such a resolution.

The following day, the bill passed the Plenary Session of the House of Councilors. The bill was put on the agenda in a meeting of the Cabinet Committee of the (Lower) House of Representatives on May 24, 2019, which seems also to be more or less a question-and-answer session, in which five politicians asked questions. The bill was enacted after passing through the Plenary Session of the House of Representatives on May 28, 2019. In the Plenary Sessions of both Houses, the vote was immediately taken after a formal report from the committee. There were neither questions nor further discussions. A similar process was seen in the enactment of the amended RTVA. The bill was put on the agenda in a meeting of the Committee on Land, Infrastructure, Transport and Tourism of both the Houses, in which mainly bureaucratic officials or the Minister of Land, Infrastructure, Transport and Tourism answered questions from member politicians from different parties within an allocated time. The bill passed the Plenary Sessions without any questions or discussions. Subsequently, some ministerial ordinances and notifications that specify details of the contents of standards and regulations, and procedures for granting conditions for road vehicles were updated by MLIT. The two laws and ordinances for the type approval came into effect in April 2020.

The question-and-answer style in the Committees, and little or no debate regarding the bills in question in the National Diet, are partly due to the absence in Japan of a thorough deliberative process for bill rewriting and discussing the bill paragraph by paragraph (called markup by the US Congress), and the lack of contestation in bill rewriting in the process [37]. Another reason is that negotiations may have occurred during unofficial processes before parliamentary deliberations. So-called "*Zizen Shinsa*" (preliminary review) by the PARC (called "*Seichōkai*" in LDP) is needed for the Cabinet to introduce bills to the National Diet. In the interaction with bureaucratic officials and interest groups, the National Diet members of the ruling party have exerted influence on the details of policy proposals [32]. Once the bill is approved by the party's general council, before it goes to the Cabinet it usually becomes subject to party discipline. In other words, once the Cabinet bill is submitted to the National Diet, politicians of the ruling party do not raise any opposition or amendment proposals. However, due to the weakened influence of "*Zoku gi'in*" by the 1994 reform and the stronger leadership exercised by the Cabinet through policy conferences, the preliminary review process was weakened. Even after the LDP's National Diet members almost lost the opportunities to influence rule-making through such unofficial processes, they remain quiet and obedient in the official sessions of the National Diet [37].

Parliamentary democracy, which weighs transparent deliberations and makes decisions that are comprehensible to the population, appears to be inactive in Japan, at least on the topic of AD. At the point of regulatory law amendments, the future mobility of/with AD is not questioned or discussed deeply by the National Diet members. Having few hurdles in rule-making will make accomplishing the initiatives much easier for the Cabinet, although it will not be achieved without the cooperation of the ministries and agencies, including their draft preparation, as can be seen from the different ministries and agencies responsible for the two laws of RTVA and RTA.

Policy conferences seem not only to work as an arena to develop strategies, but also rulemaking in recent years. The Charter for Improvement of Legal System and Environment for Automated Driving Systems was drafted in 2018 by a sub-working group, which was a subordinate meeting body of the working group (Road Traffic WG) that drafted the Public–Private ITS Initiative/Roadmaps under the IT Strategy Headquarters. The purpose was to examine what problems existed in the legal system and what kind of review was necessary to allow the driving of automated vehicles on public roads. The sub-working group was composed of technical and legal experts, as well as representatives from related ministries and agencies, and drafted the charter after four officially-recorded meetings (approximately 2 h per meeting). The achievement of a law amendment ahead of the technical development was noted as a significant success by an expert attending a meeting of the Road Traffic WG:

> A member of WG: First of all, the significance of the Public-Private ITS Initiative/Roadmaps is that the public and private sectors set a certain goal and proceed together in line with it in a concrete manner. In particular, in the area of automated driving, the development of laws usually lags behind the development of technology, but it must have been very significant that we were able to revise the laws ahead [of the technology development]. (IT Strategy Headquarters [20], p. 16)

Deregulation was counted as one of the focal measures for growth strategies to stimulate private investment under "Abenomics." Even though relaxing "*Ganban*

Kisei" (rock-solid regulations), which describes regulations that cannot be easily relaxed or eliminated due to strong opposition by those with vested interests such as government offices or interest groups, has never been easy even under top-down governance, the Cabinet has been pressurizing groups reluctant to reform. This trend continued in the Suga Cabinet (2020–2021), following the Abe administration. In a meeting of the Investment Promotion and Miscellaneous Issues Working Group under a policy conference named the Council for Promotion of Regulatory Reform in 2020, the Minister of State for Regulatory Reform urged MLIT and NPA to cooperate to relax regulations regarding field tests of AD, insisting on the economic importance of technological developments:

> Taro Kono, the Minister of State for Regulatory Reform at the time: If Japan does not lead the world in the development of automated driving, I think there will be no future for the Japanese automobile industry, but I think the reality is that there are many meaningless regulations in place, while such importance is not understood. […] If the administration does not understand that the future of the Japanese economy depends on this matter, I honestly believe that this is a big problem. I would like the National Police Agency and the Ministry of Land, Infrastructure, Transport and Tourism to always think about how important it is for the Japanese economy to create Japan's world-leading automated driving system, and the National Police Agency and the Ministry of Land, Infrastructure, Transport and Tourism to make decisions for it. (Council for Promotion of Regulatory Reform [6], p. 2)

These comments suggest that deregulation or legal reform cannot be easily implemented, even at the government's initiative, if the authority of ministries and agencies is reduced, or if there is opposition from affected interest groups. Nonetheless, responding to such pressure by the government, and following its aims for the introduction of AD transportation services at Level 4 around FY2022, and the nationwide expansion of such Level 4 transportation services in limited areas by around FY2015, a draft amendment to the Road Traffic Law to allow Level 4 automated vehicles was submitted by NPA in 2022. This was approved by the Cabinet and then passed smoothly by the National Diet. The amended law defines AD equivalent to Level 4 as "specified automated operation," and positioned it as not falling under the conventional definition of "driving." According to the newly-established "Permit System for Specified Automated Operation," business operators who wish to provide Level 4 transportation services are requested to submit a "specified automatic operation plan" and obtain prior permission from the Prefectural Public Safety Commission. This revision expanded the possibility that Level 4 AD will be introduced first for service vehicles rather than private cars. In this respect, it differs from the previous amendments to enable Level 3 AD, which allowed the Honda Legend to be the first to gain type approval and put on the market. This approach is based on the intention to realize higher Level of driving automation, starting with narrow Operational Design Domains (ODDs), where local public transportation that can be provided at limited locations within a limited time is suitable. The government expects that private vehicles will have drivers inside for a while, because, for such vehicles, technological developments that address broader ODDs are more prioritized than the achievement of higher Level of AD.

3.6 Increasing Budget

The frequent reference to AD in various policy documents and pressure for deregulation to foster the social implementation of AD indicate growing attention to the topic in the government. Accordingly, public expenses related to this topic have increased notably in recent years. The financial resources of the SIP programs that started in 2014 come from a budget allocated to the Cabinet Office, under the item named Strategic Promotion of Science and Technology Innovation Policy. 50.4 billion yen was budgeted to the item for the Cabinet Office in FY2014, which slightly increased to 56.5 billion yen in FY2020. The item was categorized under the section of Acceleration of Growth Strategies in the estimated budget requirements for FY2020, and under the section of Intensive Investment and Implementation of Digitalization as a Driving Force for Building a New Normal and Improving Productivity for FY2021. Considering that only 249 million yen was budgeted for a similar item, Promotion of Science and Technology, in 2013 for the Cabinet Office, the increase is dramatic. Within the budget, the amount of money between 2.4 and 3.5 billion yen is granted every year for the AD-related program, SIP-adus. According to the NISTEP Resource Allocation Database of the National Institute of Science and Technology Policy [34], the ratio of science and technology-related budgets by the Cabinet Office to the total of science and technology-related budgets by all the ministries increased from 0.4% in 2013 to 2.5% in 2017. That is, the attempt of the Prime Minister's Office leadership in science, technology and innovation is also evident in the budget allocations.

The budget related to AD for METI and MLIT also saw a significant increase. For FY2014, 784 million yen was allocated to the R&D and Demonstration Project of Next Generation Advanced Driver Assistance System by METI. The budget for the item, R&D and Demonstration Project Expenses for the Social Implementation of Advanced Automated Driving Systems, first appeared in FY2017 with 2.6 billion yen, increasing to 4.2 billion yen in FY2019, and to 5 billion yen in FY2020, including Mobility as a Service (MaaS) projects. As for MLIT, 339 million yen and 145 million yen were budgeted in FY2017 for the Promotion of International Standardization of Technical Standards for Automobiles and the Promotion of the Advanced Safety Vehicle Project, respectively, both of which mention AD technologies in their explanations. For FY2020, 1 billion yen was requested by MLIT for the Promotion of the Development and Commercialization of Automated Driving Technologies, in addition to 141 million yen for the Promotion of the Advanced Safety Vehicle Project. In addition, the development of standards for road space compatible with AD and support for the social implementation efforts by local governments appears to be covered under the item, Linkage of Regions and Bases through Road Networks, which was budgeted with 257.9 billion yen in total for MLIT.

Before 2013, relatively small budgets were allocated to AD, which was treated as a part of intelligent transport systems (ITS). Altogether, approximately 4.36 billion yen was allocated for the 5 year- Energy ITS Promotion Project (FY2008-2012) to the New Energy and Industrial Technology Development Organization under METI, and R&D of AD and convoy driving technologies were a part of the project. Considering that AD had previously rarely been budgeted as a stand-alone project, the current generous public expenses for the topic indicate the highest expectation of AD in the government ever.

4 Conclusion

In Japan, over the past two decades, a mechanism has been established whereby policy areas of importance to the Cabinet have been discussed in policy conferences. As evidenced by its inclusion in several policy conferences, the topic of AD has received a great deal of attention within the government in recent years. The names of the policy conferences that formulated policy documents mentioning AD give an indication of the policy areas in which AD is being addressed in Japan—foremost, the economy. Having enjoyed a fair amount of popular legitimacy in the narrative of economic revitalization in the second Abe administration (2012–2020), even IT, and science, technology and innovation policies were promoted with expectations of their contribution to economic growth and productivity improvement, and AD has been linked to these policy areas. The withdrawal of public transportation in suburban and rural areas is discussed in part with the issue of an aging society and limited social and economic activities, which is seen as a stumbling block to future economic development, and in this context, AD is expected to be a solution to the declining working population and intra-society mobility. In this way, the AD policy reflects how policymakers interpret IT, science and technology, and mobility in relation to society, and what they expect from them.

The empirical evidence on AD-related policy processes in Japan suggests that the way the government engages in the economy through various policy areas, including science, technology, and innovation, is more orchestrating than mere coordinating. The overall orchestration by the government in policy areas that are expected to enhance economic growth is reinforced by specific measures such as the promotion of AD proposed by corporate members of the technology sector. In addition to the enduring close relationship between industrial members and the government, changes in the power relationships between the Cabinet, the National Diet, and ministries and agencies have made possible the recent rather top-down policy process through policy conferences. In AD policy, the weaker legislative and judicial branches have given rise to leadership by a core executive, such as the Prime Minister, relevant Ministers of State, and other politically-appointed officials. By assigning the function of a command/control tower to the policy conferences and gathering more human resources and budgets to the Cabinet and the Cabinet Office, the strong, barely-opposed, leadership described as "*Kantei Ikkyō*" (one strong by the Prime Minister's Office) was established within the government under the Abe administration. In AD policy, such an institutional setting seems to facilitate the way of governance as an SME by reducing obstacles in the policy process for smooth policy formulation and implementation, even though ministries and agencies such as MLIT and NPA show reluctance to give up the last stronghold against deregulation.

Comparing the findings for Japan with a more general description of the German situation in this chapter indicates some significant differences between the two countries. First, AD policy and measures will be different because of varying views on how the government considers public transportation and what the government expects from scientific and technological innovations. In Germany, public transport services have been traditionally interpreted as elementary services of the public interest, and the public sector plays an important role in fulfilling its legal responsibilities in this area. In addition, German legislation mandates the incorporation of the principles of climate protection and sustainability into the provision of local public transport services. This view differs from that of the Japanese government, which has positioned public transportation as a for-profit business and, until recently, has not actively engaged in a legislative debate on the role of public transportation services and the public sector vis-à-vis society. As a result, these differences appear to affect governments' discussions on how AD should be integrated into the mobility sector, and the extent to which the governments will proactively intervene in such issues. Japanese policy seems to have a strong focus on enabling technical progress in this field. This can be seen, for instance, in that the new laws that enable Level 4 AD passed in both countries suggest its initial application in public transport. It is noteworthy that the Japanese government's motivation is strongly linked to its intention to realize a higher Level of AD technologies, even in a limited environment at first. At the same time, the German government also seems to be influencing the direction in which AD develops, with the intention of introducing it as a complement to the public transport network, and in this case at least, Germany is approaching a "state-influenced" governance style. Related to this point, the expectations of science and technology also differ to some extent between the two countries. In Germany, new scientific and technological innovations are expected to and should contribute to both economic welfare and environmental improvements. In a similar vein, in Japan, such innovations are expected to contribute to economic growth while solving social issues. In addition, however, scientific and technological independence, or technological superiority, has historically been a key concept in Japan to secure national (economic) security, regarding which policymakers share a strong sense of crisis amidst the economic stagnation of the past several decades. As a result, the state appears to intervene more directly in the coordination of scientific and technological innovation activities through state subsidies and investments in national projects to achieve the national goals. Although AD is an important policy issue in both countries in terms of industrial competitiveness because of their thriving domestic auto industries, these differences in the contextualization and expectations of AD in science and technology policy would lead to differences in approaches to investing in national innovation activities based on different underlying motivations.

It is not only differences in their motivations that shape state actions. Differences in political institutional contexts further explain whether the state can exercise its power to impose policies with relative ease, or whether the socially-embedded mechanism requires more deliberative efforts among the various stakeholders. In Germany's post-war history, social movements have had a significant influence on shaping technological innovation discourse. The ecological movements since the 1980s elevated environmental issues to a key policy agenda that policymakers can no longer ignore. Citizen participation in policy formation through the formal policy process and the public sphere, and the sensitivity of politicians to citizen input, are the result of this history and enduring attempts to enhance social and democratic legitimacy. Accordingly, the German government tends to play the role of mediator in social negotiations by enabling exchanges among citizens, science, and industry. The survey results presented in chapter "Social Acceptance of CAD in Japan and Germany: Conceptual Issues and Empirical Insights" suggest the comparatively high willingness of the German public for involvement in planning and conducting AD field trials, which may be an indication of their interest in the governance of emerging technologies in the pre-market stage. Furthermore, German policymakers are required to formulate relevant policies at multi-levels of governance—at the European, national, and local levels. Responsibility for the planning, organization, and financing of local public transport is decentralized, and the Federal Government needs to function as a policy moderator between the European Union and regional/local communities.

As in Japan, the German technology sector, especially the automotive industry, has a close relationship with the government and channels to influence the policy direction of AD. Yet, perhaps in Japan, as long as the industry proposals support the government's overall leadership in science and technology development, the players may be able to influence policy in a more overt and visible way, as evidenced by industry participation in policy conferences. In addition, in Japan, civil society is hardly represented in AD-related policy processes when looking at members of relevant policy conferences. Government actions seem to be largely legitimated by democratic electoral-based state legitimacy, in which economic policy has been a key election issue for many years. Having enjoyed a fair amount of popular legitimacy in the narrative of economic revival, the second Abe administration (2012–2020), together with industry, successfully linked AD to the expected positive consequences of national competitiveness and productivity improvement. While citizens seldom show doubt about the paternalistic leadership role of the government in economic growth, the government expects the general public to improve their knowledge and understanding of the use of AD through communication activities and, in turn, become responsible consumers/users of new products and services. Such political institutions and governance modes may mean that, compared to Germany, Japan can promote AD policies relatively smoothly through means such as investment in national projects and modification of the legal framework in the name of governmental leadership supported by citizens and industry. Therefore, if future cabinets continue to place AD at the center of their policy interests, the introduction of automated vehicles to the market may even be realized sooner than in other countries. However, just as the privatization of public transportation has led to a rapid decline in services in rural and suburban areas, the economic-oriented promotion of AD will not necessarily guarantee a sustainable public transportation system. On the other hand, Germany is ahead of other countries in legal amendments related to AD. However, policymakers are required to play a role in the multi-level governance participated by diverse stakeholders, and the social implementation of AD will proceed through a responsible governance approach, which does not promise the speedy realization of AD in German society. Nevertheless, for a country seeking to achieve public value by enhancing the democratic legitimacy of its policy decisions in the future sociotechnical system, this approach may be a shorter path to this end.

In this chapter, we put the focus on a detailed analysis of the Japanese situation in the context of the governance of AD. On a more general level, we made some comparisons with the situation in Germany. On that basis, the analysis in this chapter indicates that the broader settings of technology governance, here framed as politics, polity, and policy, may well have a significant influence on the future development of AD. Automated vehicles come with an immense transformative potential for the entire mobility sector [5, 44]. To allow effective and efficient governance in a complex field such as mobility, we need a good understanding of all factors that may influence future development pathways. Further research should take up these findings and undertake more comparative analyses between countries that have the potential to play a leading role in future governance and implementation of AD.

Appendix

Table 1 Japanese policy conferences, ministries and agencies that have published policy documents mentioning automated driving (AD) between 2012 and 2020

Policy conference/Ministry and agency	Member of conference	Subordinate meeting body engaged in draft writing	Member of meeting body	Name of policy document that mentions automated driving (AD)
Headquarters for Japan's Economic Revitalization (2012–2020 in the Cabinet) Committee on the Growth Strategy (since 2020 under the Cabinet)	Head: PM Acting head: Deputy PM/MoF Deputy head: CCS; MoSEF Member: other Ministers (as of 16.09.2020)	Council on Investments for the Future	Head: PM Acting head: Deputy PM/MoF Deputy head: CCS; MoETI; MoSEF Member: MoECSST; MoIAC; MoSRR; MoSSTP; 5 from business field (DeNA, Future Corp.; Japan Business Federation; Nissan; Sompo Holdings, etc.); 2 from univ. (as of 05.10.2018)	Growth Strategy (2013), revised and published every year
Council on Overcoming Population Decline and Vitalizing Local Economy (since 2014 under the Cabinet)	Head: PM Deputy head: CCS Member: other Ministers; 4 from business field (East Japan Railway Company, etc.); 3 from univ.; 3 from other fields (as of 01.05.2020)			Basic Policy for Overcoming Population Decline and Vitalizing Local Economy (2015), revised and published every year

(continued)

Table 1 (continued)

Policy conference/Ministry and agency	Member of conference	Subordinate meeting body engaged in draft writing	Member of meeting body	Name of policy document that mentions automated driving (AD)
Integrated Innovation Strategy Promotion Council (since 2018 under the Cabinet)	Head: CCS Acting head: MoSSTP Deputy head: MoITP; MoSHP; MoSIPS; MoSOP; MoSSP Member: other Ministers (as of 25.07.2018)			Integrated Innovation Strategy (2018), revised and published every year
IT Strategy Headquarters (since 2001 in the Cabinet)	Head: PM Deputy head: CCS; MoETI; MoIAC; MoITP Member: other Ministers; Government CIO; 6 from business field (Japan Business Federation; NTT; Preferred Networks; Rakuten, etc.); 3 from univ.; 1 from other fields (as of 19.12.2018)	IT Strategy Drafting Committee	Chairman: Government CIO Member: 7 from business field (Future Corp.; NTT; Toyota, etc.); 1 from univ.; 1 from other fields (as of 09.05.2013)	Declaration to be the World's Most Advanced IT Nation (2013), revised and published every year until 2016
		Executive Committee of Basic Act on the Advancement of Public and Private Sector Data Utilization	Chairperson: Prof. from univ. Member: 7 from business field (Hitachi; NTT Data; Rakuten; Toyota, etc.); 6 from univ.; 4 from other fields; Government officials from different ministries and agencies (as of 04.04.2017)	Declaration to Be the World's Most Advanced IT Nation Basic Plan for the Advancement of Public and Private Sector Data Utilization (2017), revised and published every year until 2020

(continued)

Table 1 (continued)

Policy conference/Ministry and agency	Member of conference	Subordinate meeting body engaged in draft writing	Member of meeting body	Name of policy document that mentions automated driving (AD)
		Road Traffic Working Group	Chairperson: Prof. from univ. Member: 4 from business field (DeNA; Hitachi Automotive Systems; Japan Automobile Manufacturers Association; Toyota); 2 from univ.; 3 from other fields; Government officials from different ministries and agencies (as of 05.12.2018)	Public–Private ITS Initiative/ Roadmaps (2014), revised and published every year
		Sub-Working Group for the Charter for Improvement of Legal System and Environment for Automated Driving Systems	Chairperson: Prof. from univ Member: 4 from business field (DeNA; Hitachi Automotive Systems; Japan Automobile Manufacturers Association; Toyota); 5 from univ.; 3 from other fields; Government officials from different ministries and agencies (as of 07.11.2018)	Charter for Improvement of Legal System and Environment for Automated Driving Systems (2018)

(continued)

Table 1 (continued)

Policy conference/Ministry and agency	Member of conference	Subordinate meeting body engaged in draft writing	Member of meeting body	Name of policy document that mentions automated driving (AD)
Council for Science, Technology and Innovation (since 2014 in the Cabinet Office)	Head: PM Member: CCS; MoECSST; MoETI; MoF; MoIAC; MoSSTP; Science Council of Japan; 3 from business field (Toyota, etc.); 4 from univ. (as of 28.02.2018)	Study group on how to promote fundamental technologies	Member: Science Council of Japan; 7 from business field (Council on Competitiveness-Nippon; Hitachi; Toyota, etc.); 14 from univ. (04.09.2015)	The Science and Technology Basic Plan, revised and published every five year
		SIP-adus Steering Committee	Program Director: Toyota Deputy Program Director: Nissan; Honda; Japan Science and Technology Agency; Prof. from univ. Member: 6 from business field (Honda; Japan Auto Parts Industries Association; Japan Automobile Standards Internationalization Center; Japan Electronics and Information Technology Industries Association; Japan Federation of Hire-Taxi Associations; Suzuki, etc.); 5 from univ.; 3 from other fields; Government officials from different ministries and agencies (as of 05.06.2019)	R&D Plan for SIP-adus (2014), revised and published every year

(continued)

Table 1 (continued)

Policy conference/Ministry and agency	Member of conference	Subordinate meeting body engaged in draft writing	Member of meeting body	Name of policy document that mentions automated driving (AD)
Ministry of Economy, Trade and Industry; Ministry of Land, Infrastructure, Transport and Tourism		Panel on Business Strategies for Automated Driving	Chairperson: Prof. from univ. Member: 14 from business field (Toyota; Nissan; Honda; Isuzu; Hino; Denso; Jtect; Renesas electronics; Hitachi Automotive Systems; Panasonic, etc.); 5 from univ.; 3 from other fields (as of 31.03.2020)	"Action Plan for realizing the Automated Driving" Version 4.0 by Panel on Business Strategies for Automated Driving (2016), revised and published every year
National Police Agency		Research and Study Committee for the Realization of Automated Driving	Member: 1 from business field (Japan Automobile Manufacturers Association); 9 from univ.; 2 from other fields; Government officials from National Police Agency (as of 06.11.2020)	Report on Research and Studies for the Realization of Automated Driving (2020), revised and published every year

Note Document name in bold means that the document was brought up and approved in a Cabinet meeting
List of abbreviations:
CCS: Chief Cabinet Secretary
MoECSST: Minister of Education, Culture, Sports, Science and Technology
MoETI: Minister of Economy, Trade and Industry
MoF: Minister of Finance
MoIAC: Minister of Internal Affairs and Communications
MoITP: Minister in charge of Information Technology Policy
MoSEF: Minister of State for Economic and Fiscal Policy
MoSHP: Minister of State for Healthcare Policy
MoSIPS: Minister of State for the Intellectual Property Strategy
MoSOP: Minister of State for Ocean Policy
MoSRR: Minister of State for Regulatory Reform
MoSSP: Minister of State for Space Policy
MoSSTP: Minister of State for Science and Technology Policy
PM: Prime Minister
Univ.: university or research institute

References

1. Anchordoguy, M. (2005). *Reprogramming Japan: The high tech crisis under communitarian capitalism*. Cornell University Press.
2. Brandt, T. (2006). Liberalisation, privatisation and regulation in the German local public transport sector. Country reports on liberalisation and privatisation processes and forms of regulation. *Wirtschafts- Und Sozialwissenschaftliches Institut Der Hans Böckler Stiftung*. https://www.boeckler.de/pdf/wsi_pj_piq_impact_liberalisation.pdf.
3. Burrett, T. (2017). Abe road: Comparing Japanese Prime Minister Shinzo Abe's leadership of his first and second governments. *Parliamentary Affairs, 70*(2), 400–429. https://doi.org/10.1093/pa/gsw015.
4. Burri, R. V. (2015). Imaginaries of Science and Society: Framing Nanotechnology Governance in Germany and the United States. In S. Jasanoff & S.-H. Kim (Eds.), *Dreamscapes of Modernity Sociotechnical Imaginaries and the Fabrication of Power*. The University of Chicago Press.
5. Canzler, W., & Knie, A. (2016). Mobility in the age of digital modernity: why the private car is losing its significance, intermodal transport is winning and why digitalisation is the key. *Applied Mobilities, 1*, 56–67. https://doi.org/10.1080/23800127.2016.1147781.
6. Council for Promotion of Regulatory Reform. (2020). *Minutes of the 6th meeting of the Investment Promotion and Miscellaneous Issues Working Group*. https://www8.cao.go.jp/kisei-kaikaku/kisei/meeting/wg/toushi/20201215/gijiroku1215.pdf (in Japanese, accessed on August 25. 2021).
7. Docherty, I., Marsden, G., & Anable, J. (2018). The governance of smart mobility. *Transportation Research Part A: Policy and Practice, 115*, 114–125. https://doi.org/10.1016/j.tra.2017.09.012.
8. Ebner, A. (2016). Institutional Transformations of Technology Policy in East Asia: The Rise of the Entrepreneurial State. In U. Hilpert (Ed.), *Routledge Handbook of Politics and Technology* (1st ed., pp. 369–379). Routledge.
9. Fujimura, N. (2012). Electoral incentives, party discipline, and legislative organization: Manipulating legislative committees to win elections and maintain party unity. *European Political Science Review, 4*(2), 147–175. https://doi.org/10.1017/S1755773911000166.
10. Fukukawa, S., Amatori, F., & Molteni, C. (2017). Industrial Policy and the Role of MITI in Japan. In D. Felisini (Ed.), *Reassessing the Role of Management in the Golden Age: An International Comparison of Public Sector Managers 1945–1975* (pp. 141–152).
11. Gerstenfeld, A. (1982). *Science Policy Perspectives: USA–Japan*. Academic Press, Inc.
12. Government of Japan. (2017). *New Economic Policy Package*. https://www5.cao.go.jp/keizai1/package/20171208_package_en.pdf (accessed on August 25. 2021).
13. Government of Japan. (2018). *Integrated Innovation Strategy*. https://www8.cao.go.jp/cstp/english/doc/integrated_main.pdf (accessed on August 25. 2021).
14. Government of Japan. (2020). *Integrated Innovation Strategy 2020*. https://www8.cao.go.jp/cstp/english/strategy_2020.pdf (accessed on August 25. 2021).
15. Hall, P. A., & Soskice, D. (2001). *Varieties of capitalism: The institutional foundations of comparative advantage*. Oxford University Press.
16. Holliday, I. (2000). Productivist welfare capitalism: Social policy in East Asia. *Political Studies, 48*(4), 706–723. https://doi.org/10.1111/1467-9248.00279
17. Hundt, D., & Uttam, J. (2017). Japan's Collective Capitalism and the Origins of the Asian Model. In *Varieties of Capitalism in Asia* (pp. 39–76). Palgrave Macmillan.
18. IT Strategy Headquarters. (2013a). *Declaration to be the World's Most Advanced IT Nation*. https://japan.kantei.go.jp/policy/it/2013/0614_declaration.pdf (accessed on August 25. 2021).
19. IT Strategy Headquarters. (2013b). *Minutes of the 60th meeting of the IT Strategy Headquarters*. https://www.kantei.go.jp/jp/singi/it2/dai60/gijiroku.pdf (in Japanese, accessed on August 25. 2021).
20. IT Strategy Headquarters. (2019). *Minutes of the 6th meeting by the Road Traffic Working Group*. https://www.kantei.go.jp/jp/singi/it2/dourokoutsu_wg/dai6/gijiyousi.pdf (in Japanese, accessed on August 25. 2021).

21. IT Strategy Headquarters. (2020). *Kanmin ITS Kōsō Rōdomappu 2020 [Public-Private ITS Initiative/Roadmaps 2020]*. https://www.kantei.go.jp/jp/singi/it2/kettei/pdf/20200715/2020_r oadmap.pdf.
22. Ibata-Arens, K. C. (2019). *Beyond Technonationalism: Biomedical Innovation and Entrepreneurship in Asia*. Stanford University Press. https://doi.org/10.1515/9781503608757
23. Kagawa, M. (2012). Rūraruchiiki ni okeru kōkyōkōtsū no iji • saisei to kōtsūkihonhōan (in Japanese) [Maintenance and Revitalization of Public Transport in Rural Areas and Basic Transportation Bill]. *The Economic Review of Kumamoto Gakuen University, 18*(3•4), 1–26.
24. Karo, E. (2018). Mission-oriented innovation policies and bureaucracies in East Asia. *Industrial and Corporate Change, 27*(5), 867–881. https://doi.org/10.1093/icc/dty031.
25. Kobayashi, S., Akaike, S., Hayashi, T., Tomizawa, H., & Miyabayashi, M. (2019). Kagaku Gijutsu Kihonkeikaku no Henkan to Jiki eno Tembō (in Japanese) [Changes in the Science and Technology Basic Plan and Prospects for the Next Phase]. *The Journal of Science Policy and Research Management, 34*(3), 190–215. https://doi.org/10.20801/jsrpim.34.3_190.
26. Kohno, M. (2002). A changing Ministry of International Trade and Industry. In J. Amyx & P. Drysdale (Eds.), *Japanese Governance: Beyond Japan Inc.* (1st ed.). Routledge.
27. Kubo, H. (2016). Naikaku no shudō ni yoru shōrai no seisaku mokuhyō no kettei to senmonteki chiken no yakuwari (in Japanese) [Determining future policy goals at the initiative of the Cabinet and the role of expertise]. *Kōnan Hōgaku, 56*(3•4), 163–202.
28. Machidori, S. (2015). Kantei kenryoku no hen'yō. Shushō dōkō dēta no hōkatsuteki bunseki o tegakarini (in Japanese) [The Transformation of the Japanese Prime Ministerial Power: A Comprehensive Analysis of the Prime Minister's Meeting Records]. *Senkyo Kenkyū, 31*(2), 19–31.
29. Maruyama, M. (2013). "Being" and "Doing" (1958) (D. Washburn, Trans.). *Review of Japanese Culture and Society, 25*, 152–169. https://www.jstor.org/stable/43945391.
30. Matsui, S. (2011). Why is the Japanese Supreme Court so Conservative? *Washington University Law Review, 88*(6), 1375–1423. https://openscholarship.wustl.edu/law_lawreview/vol88/iss6/2/.
31. Merkel, W. (2004). Embedded and defective democracies. *Democratization, 11*(5), 33–58. https://doi.org/10.1080/13510340412331304598.
32. Mishima, K. (2019). The Presidentialization of Japan's LDP Politics: Analyzing Its Causes, Limits, and Perils. *World Affairs, 182*(1), 97–123. https://doi.org/10.1177/0043820019826671.
33. Mogaki, M. (2019). *Understanding governance in contemporary Japan*. Manchester University Press.
34. National Institute of Science and Technology Policy. (n.d.). *NISTEP Resource Allocation Database*. http://www.nistep.go.jp/research/scisip/data-and-information-infrastructure (accessed on August 25. 2021).
35. Nikkei. (2020, September 18). *Keisankanryō ni huyu no yokan. Sinseiken, syusyōhosakan ra kōtai*. (in Japanese) *[Feeling of winter in the METI bureaucracy. New administration, replacing special advisor to the prime minister.]*.
36. Nikkei. (2021, July 27). *Kyūma Kazuo shi (21). Sosaetei 5.0 teishō*. (in Japanese) *[Mr. Kyūma Kazuo (21). Proposal of Society 5.0]*.
37. Nonaka, N., & Aoki, H. (2016). *Seisaku kaigi to tōron naki kokkai. Kantei shudō taisei no seiritsu to kōtai suru jukugi* (in Japanese) *[Policy councils and the Diet without debate: The Establishment of the executive-led system and the retreat of deliberation]*. Asahi Shimbun Publications Inc.
38. Odagiri, H. (2007). The national innovation system: a key to Japan's future growth. In D. Bailey, D. Coffey, & P. Tomlinson (Eds.), *Crisis or Recovery in Japan: State and Industrial Economy* (pp. 157–178). Edward Elger Publishing Limited.
39. Pempel, T. J. (2012). Between Pork and Productivity: The Collapse of the Liberal Democratic Party. In M. Ido (Ed.), *Varieties of Capitalism, Types of Democracy and Globalization*. Routledge.
40. Rotmans, J., Kemp, R., & Asselt, M. van. (2001). More evolution than revolution. *Foresight, 3*(1), 15–31. https://doi.org/10.1108/14636680110803003.

41. Sakai, K. (2008). Comparison Study of the Urban Transportation Planning Systems in Major Developed Countries - Towards the Better Urban Transportation Planning System through the Analysis on those of France, USA, Germany, UK, and Japan - (in Japanese). *Journal of the City Planning Institute of Japan, 43*(3), 937–942.
42. Saruwatari, T. (2020). Gyōsei tono Kankeisei ni Motozuku NPO no Soshiki Ruikei (in Japanese) [Organizational Types of NPOs Based on the Relationship with Government]. *Social Analysis, 47,* 81–94.
43. Schaede, U. (1995). Positive Regulierung—Staat und Unternehmen im japanischen Wirtschaftswachstum. In G. Foljanty-Jost & A.-M. Thränhardt (Eds.), *Der schlanke japanische Staat* (pp. 106–121). Leske + Budrich.
44. Schippl, J., Truffer, B., & Fleischer, T. (2022). Potential impacts of institutional dynamics on the development of automated vehicles: Towards sustainable mobility? *Transportation Research Interdisciplinary Perspectives, 14,* Art.-Nr.: 100587. https://doi.org/10.1016/j.trip.2022.100587.
45. Schmidt, V. A. (2007). Bringing the State Back Into the Varieties of Capitalism And Discourse Back Into the Explanation of Change. *Center for European Studies, Program for the Study of Germany and Europe Working Paper Series, 07.3.*
46. Setoyama, J. (2015). Naikakukambō ● naikakufu no gyōmu no surimu ka (in Japanese) [Slimming Down the Operations of the Cabinet Secretariat and Cabinet Office]. *Rippō to Chōsa, House of Councillors, 364,* 3–17.
47. Takahashi, Y. (2007). Provision of local bus services in Japan: focusing on the roles for local governments and nonprofit organisations. *Proceedings of the 10th International Conference on Competition and Ownership in Land Passenger Transport (THREDBO 10).* https://ses.lib rary.usyd.edu.au/handle/2123/6052.
48. Takano, Y., & Taniguchi, M. (2018). A Study on the government expenditure for public transportation policy by municipalities - An analysis of the questionnaire survey targeting all cities in Japan - (in Japanese). *Journal of the City Planning Institute of Japan, 53*(3), 1385–1392.
49. Takenaka, H. (2021, May 21). Gokuhi Kaigō "CPU." Seikenchūsū no shireitō. Sōrishudō ha kore de jitsugen shita (in Japanese) [Top secret meeting "CPU." The command tower of inner circle of the administration. The Prime Minister's leadership was realized by this]. *[Video]. YouTube.* https://www.youtube.com/watch?v=vX59FRU3hs0&list=PLV8yIIai M8u14PURMKHTO4kEpsfJSGc8N&index=11 (accessed on August 25. 2021).
50. Tanaka, H. (2019). Dai niji abe seiken ni okeru seisakukeiseikatei no gabanansu -konsutabiritei no shiten kara- (in Japanese) [Governance of the Policy Making Process in the Second Abe Administration: From the Perspective of Contesterability]. *Nenpō Gyōsei Kenkyū, 54,* 57–82.
51. Tanaka, Y., Chapman, A., Tezuka, T., & Sakurai, S. (2020). Multiple Streams and Power Sector Policy Change: Evidence from the Feed-in Tariff Policy Process in Japan. *Politics and Policy, 48*(3), 464–489. https://doi.org/10.1111/polp.12357.
52. Taniguchi, T. (2018). *Abe Shinzo no shinjitsu* (in Japanese) *[The Truth about Shinzo Abe].* Kabushiki Kaisha Gokū Shuppan.
53. Truffer, B., Schippl, J., & Fleischer, T. (2017). Decentering technology in technology assessment: prospects for socio-technical transitions in electric mobility in Germany. *Technological Forecasting and Social Change, 122,* 34–48. https://doi.org/10.1016/j.techfore.2017.04.020.
54. Tsugawa, S. (2008). A history of automated highway systems in Japan and future issues. *Proceedings of the 2008 IEEE International Conference on Vehicular Electronics and Safety,* 2–3. https://doi.org/10.1109/ICVES.2008.4640914.
55. Vogel, S. K. (2001). THE CRISIS OF GERMAN AND JAPANESE CAPITALISM: Stalled on the Road to the Liberal Market Model? *Comparative Political Studies, 34*(10), 1103–1133.
56. Wakabayashi, K., Griffy-brown, C., & Watanabe, C. (1999). Stimulating R&D: an analysis of the Ministry of International Trade and Industry's 'visions' and the current challenges facing Japan's technology policy-making mechanisms. *Science and Public Policy, 26*(1), 2–16.
57. Watari, N., & Kobayashi, Y. (2020). Chōkiseiken wo Kanō tosuru Jōken ha Nanika?: Chōkiseiken to Tankiseiken no Hikakubunseki (in Japanese) [What is Necessary Conditions for Establishing Long-term Government?: A Comparative Analysis between Long-term and

Short-lived Government]. *The Research Bulletin of Iryo Sosei University -Humanities, Social Sciences and Informatics-, 33*(5), 44–59.

58. Weiss, L. (2009). Guiding globalisation in East Asia: new roles for old developmental states. *States in the Global Economy*, 245–270. https://doi.org/10.1017/cbo9780511491757.013.

59. Witt, M. A., & Redding, G. (2009). Culture, meaning, and institutions: Executive rationale in Germany and Japan. *Journal of International Business Studies, 40*(5), 859–885. https://doi. org/10.1057/jibs.2008.81.

60. Yamamura, K., & Streeck, W. (2003). *The End of Diversity? Prospects for German and Japanese Capitalism*. Cornell University Press.

Business Analysis and Prognosis Regarding the Shared Autonomous Vehicle Market in Germany

Jonas Hennig, Thomas Meissner, Marcus Krieg, and Felix Przioda

Abstract Driven by the current advancements in autonomous driving (AD) technology, operations of shared autonomous vehicles (SAVs) are increasingly moved from closed test sites to public roads worldwide. Since the market potential of SAV services in Germany remains largely unexplored, this chapter aims to provide a holistic business analysis and prognosis of the German SAV market. To adequately capture the operational complexity of future SAV fleets, a simulation-based approach is used. The simulation model is created utilizing proprietary operational performance and cost data of actual shared vehicle services and pilots in Germany. Three market demand scenarios are simulated and combined with a detailed cost scenario for the year 2035. Results indicate improved operational efficiency indicators for SAV fleets compared to non-autonomous fleets. However, we also found that cost structures of SAV fleets may be highly underestimated in previous studies, resulting in a per VKM cost range of 0.48–0.50 EUR. While customer price levels are significantly lower than the current cost of using ride-hailing services, they may not be low enough to convince a considerable number of passengers to switch from their private cars to SAV services. The results underscore the importance of collaborative efforts between policymakers, fleet operators and manufacturers to harness the benefits of SAV fleets.

J. Hennig (✉) · T. Meissner · M. Krieg · F. Przioda
BMW Group, Munich, Germany
e-mail: Jonas.Hennig@partner.bmwgroup.com

T. Meissner
e-mail: thomas.r.meissner@bmwgroup.com

M. Krieg
e-mail: Marcus.Krieg@bmw.com

F. Przioda
e-mail: Felix.Przioda@bmw.de

© The Author(s) 2025
C. Eisenmann et al. (eds.), *Acceptance and Diffusion of Connected and Automated Driving in Japan and Germany*, https://doi.org/10.1007/978-3-031-59876-0_4

1 Introduction

In September 2015, legally blind Steve Mahan was the passenger of the world's first fully autonomous trip on public roads, conducted by the Google self-driving car project in Austin, USA [18]. Public testing of shared autonomous vehicles (SAVs) started before 2010, but this was the first time a trip took place on public roads, without a police escort or safety personnel onboard. After early hype and successes followed by a phase of delayed milestone achievements and unsuccessful pilots, SAVs now seem to be on the way to becoming stable operational mobility services. In fact, Waymo One, the first commercial SAV service, started operating in October 2020 in Phoenix, USA. SAV services are considered to have a high potential to relieve especially urban mobility systems, which are faced with a growing volume of traffic. Congested roads and parking space pressure are particularly noticeable at peak times in large cities. As the share of the global population living in cities is expected to grow to 68% by 2050, the overload of mobility systems will only increase without innovative mobility options [35, 42].[1] SAV services offer the possibility of transporting people by sharing journeys or vehicles without each person needing their own vehicle, and thus offer an opportunity to relieve urban congestion. Moreover, SAVs are regarded as the key to profitability for mobility service businesses, since the highest cost element—the manual driving task—is no longer needed. Thus, due to the current advancements in autonomous driving technology and actual roll-out of the first SAV services, the global ride-hailing market is expected to grow from 130 billion Euro revenues in 2021 to 1.2–1.6 trillion Euro by 2030 [16, 30]. Furthermore, based on the global investment in mobility sector start-ups since 2009, providers of ride-hailing and carsharing services attracted 24.8 billion Euros of investments [16].

The positive market projections, however, mostly assume constant passenger fares over time. This may not be the case considering the ongoing high competition in the current ride-hailing and carsharing market. Even if cost savings can be realized by eliminating the manual driving task, it stands to reason that competition will only be continued at a lower price level. Moreover, recent research on urban archetypes shows that not every city may benefit from autonomous vehicle (AV) mobility [26]. Based on these insights and learning over the past decade, policymakers in urban areas have developed a more differentiated view of the advantages and disadvantages of SAV services [2, 17, 21]. European cities especially are already taking measures to ensure that mobility services complement rather than substitute the public transport network, avoiding a radical shift from mass- to more private transport. This may considerably limit the market development of SAV services. Furthermore, regarding the underlying cost structure, studies have mainly focused on the comparison of privately owned transportation and SAV services. While this has contributed to understanding the impact if people switch from private cars to shared vehicles, it does not necessarily mean that the mobility services will operate profitably. To date, profitability analyses

---

[1] Please note that these studies were published before the global COVID-19 pandemic which may influence the stated projections.

of SAV fleet providers based on actual operational cost data are scarce, or mostly rely on estimates, owing to insufficient data regarding underlying cost structures.

Hence, the objective of this study is to provide a holistic business analysis and prognosis of the SAV service market in Germany. We selected Germany as it represents a mature mobility market and thus offers a reliable data basis. Moreover, as the focus of past SAV research was mainly on cities in the USA, we believe it is beneficial to take a closer view of the shared vehicle mobility system in urban areas in Germany, which significantly differs. Three lead questions guide this study:

1. What is the current demand for shared vehicle services in Germany and how may it develop until 2035?
2. What are the operational implications for an SAV service provider to meet the future demand?
3. What does that mean for the provider's cost structure, and which customer price level is needed for a profitable business operation in 2035?

To assess the current and future German shared vehicle market and the respective business implications from an operator's perspective, the following approach is taken: as a first step, this study is intended to analyze the current shared vehicle service market and derive possible future scenarios as to how the market may evolve with the introduction of AVs. Secondly, it is examined how a potential SAV fleet operator can reasonably serve the future demand. Finally, the study analyzes if a fleet operator can run an SAV fleet profitably, by taking operational cost parameters into account. The underlying cost structure is derived from actual operational data of shared vehicle services and pilots.

2 Review

2.1 Classification of Shared Autonomous Vehicle Services

In practice, the distinction between the different autonomous mobility services is not trivial. Classifications vary internationally and neither academia nor the different service providers have agreed on one common use of language. At vehicle level, however, there is a high degree of consensus in theory and practice on the distinction between privately owned AVs and shared, commercially operated AVs (see Fig. 1; Bösch et al. [6]). The dividing line between private and shared vehicles may become blurred when privately owned AVs are temporarily made accessible to serve in a shared fleet, similar to the Airbnb approach on the accommodation market.

Within the shared, commercial operation of AVs, a relevant distinction should be made based on the privacy character of the transport. One option is that passengers are transported privately to their destination without unnecessary detours ($SAV_{private}$). If several passengers with similar departure and destination points are transported, then the service is classified as pooled (SAV_{pooled}) which implies that passengers

Fig. 1 Classification of AVs based on vehicle use and transport privacy[2]

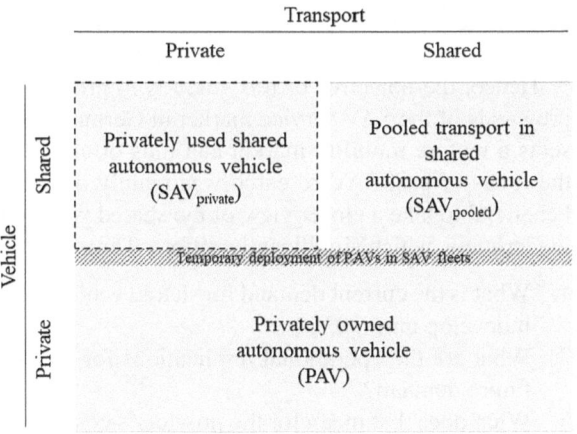

are willing to accept minor deviations on the route to pick up or drop off other passengers.[3] The focus of this study is on $SAV_{private}$ services, as indicated in Fig. 1 (dashed line). An important assumption of this study is that the classification of an SAV service as private transport does not necessarily mean exclusive transport of one person: it includes the transport of several people who know each other (e.g., friends or family) who have the same departure and destination points for a journey. Consequently, occupancy rates of $SAV_{private}$ can be >1.

This study concentrates on the Society of Automotive Engineers (SAE) automation Levels 4 (High Automation) and 5 (Full Automation) (for levels of AD see the introduction). The term 'autonomous' is used for both automation levels throughout this chapter.

2.2 Market Demand Models and Prognoses of Shared Autonomous Vehicle Services

In recent years, established management consultancies have forecast an enormous growth in shared vehicle services, such as ride-hailing and carsharing services, and based on the recent advancements of the AV technology, continue to do so [13, 15, 23, 25]. However, despite strong growth in all types of services, to date actual growth has not been achieved to the forecasted extent. Considering the academic literature, researchers generally agree on the potential of shared vehicle services to influence the travel behavior of different groups of people and thus to influence the respective passenger kilometers (PKM) and vehicle kilometers (VKM, in the USA mainly

---

[2] Modified illustration based on Richter [34].

[3] The terms ride-pooling and ride-sharing are not sufficiently delimited and cause confusion in practice. This study adheres to a widely accepted distinction in the transport industry, which assigns ride-pooling to commercial use and ride-sharing to private use.

referred to as Vehicle Miles Travelled, VMT) covered [28]. Studies of the current market demand for shared vehicle services have mainly concentrated on so-called Transportation Network Companies (TNCs) in US cities. In 2019 for example, the share of TNCs of all VMTs was 12.8% in central San Francisco and 2.6% in central Los Angeles [3]. In New York City, the accumulated vehicle kilometers of TNCs, traditional taxis and limousine chauffeur services increased from 14% in 2013 to 19% of the total VKM in 2016 [36]. Further research in Boston showed a VMT share of 3.9% for the inner city, which equates to 291 million VMTs in 2018 [31]. The results show that the penetration of shared vehicle services differs even across cities in the same geographical region. To the authors' knowledge, comparable research results for German cities is still missing. Given the current focus on the USA as a preferred study region, a transfer of available research results to the German market should not be undertaken without restrictions. Firstly, TNCs are still heavily regulated in many European countries. Moreover, European cities usually have a much better developed public transport system compared to the USA study areas. Thus, a rapid increase in market penetration as seen in American cities is not to be expected.

As most studies in this context examine the interactions and shifting effects to and from traditional means of transport, additional VKM created by TNCs or their vehicle-related emissions, VKMs is selected as the central reference value for the analysis of market demand [12]. However, to derive the market demand, the analysis should be conducted on the individual passenger level (PKM), as two or more passengers may share the same vehicle. An increase in VKM while PKM remain constant may be due to inefficiencies of the service, such as empty runs or low utilization in general. In the model of [27], the AV service improved vehicle utilization and showed average occupancy rates ranging from 1.4 to 1.7 persons across their scenarios.

Looking ahead, researchers showed that especially AV technology has the potential to significantly accelerate the growth and integration of shared vehicle services, alongside the technological development of mobility apps, data analysis and artificial intelligence [8, 41]. Further market diffusion of AV technology would allow for different types of driver-dependent services, such as carsharing, ride-pooling and ride-hailing, to merge seamlessly. Since all types of services can be provided by AVs, there would be no reason for the operator to let the customers drive the vehicle themselves. In addition, most researchers agree that AV technology enables providers to offer their services at lower prices, which further increases their attractiveness [37].

However, decreasing customer prices may have not only technological but also regulatory limits, as cities may try to avoid substitution effects from public transport to motorized private transport through tolls, or limitation of operating areas or operating hours. Cities are already taking measures to ensure that shared vehicle services complement rather than substitute the public transport network. This may considerably limit the market development of shared vehicle services. Moreover, recent research on urban archetypes shows that not every city may benefit from the introduction of SAV services [26]. Based on these insights and learning over the past decade, policymakers in urban areas have developed a more differentiated view of the advantages and disadvantages of SAV services [2, 17, 21]. This is reflected in the new German Autonomous Driving Act, which became effective in July 2021: The regulated operational scenarios particularly include demand-oriented SAV services in off-peak times and the transport of people on the first or last mile of their journey [7]. Nevertheless, the Autonomous Driving Act as a nationwide legal framework enables AVs to drive in pre-defined operating areas on public roads in regular operation. As a first reaction to the new regulatory framework, AV technology provider Mobileye and fleet operator Sixt are planning to start a robotaxi pilot service in Munich by the end of 2022 [22].

2.3 Cost Structure Analyses and Prognoses of Shared Autonomous Vehicle Services

Overall, researchers assume that AVs will be used initially in shared vehicle service fleets before being available for the private customer. This is mainly due to the fact that AV technology, in its current state without benefiting from scaling effects, is more expensive than private customers would be willing to pay. Specific AV technology cost estimates vary considerably: while Jones and Leibowicz [24] estimate costs of 7,500–10,000 USD, Wadud [43] suggests a range of 13,700–16,600 USD for the current state of the technology. Cost indications of current AV technology providers range from 10,000–36,000 USD [14, 40]. However, all cost estimates must be treated cautiously, as it remains unclear if studies present purchase prices (including profits), or cost prices. Furthermore, studies partly rely on rough estimates which are not comprehensively validated by industry experts. Notwithstanding the considerable variation, all cost ranges exceed the private customer's willingness to pay (WTP) for the addition of an autonomous system: based on a customer survey in the USA, the average WTP is 7,253 USD [4].

Commercial shared vehicle service providers, however, may still benefit from the use of AV technology. Current research suggests that AV technology presents a high cost-saving potential for fleet operators, as they may offer the same conventional ride-hailing or taxi service while saving the cost of the manual driving task [43]. Bösch et al. [6] predict that the realization of higher utilization of vehicles may lead to competitive customer prices as low as the prices for public transport. Moreover, in a long-term view with a high number of AVs on the market, technology advancements and scale effects may lead to AV system costs of 1,000–5,000 USD [9, 24]. Amnon Shashua, CEO of AV technology provider Mobileye, estimated their system cost will reach the 5,000 USD target as soon as 2025 [40].

In addition to the AV system costs, an autonomous ride-hailing provider must cover various operational costs. Researchers have identified several parameters which are relevant for service providers and may certainly be influenced by AV technology:

- Vehicle lifetime [1, 27]
- Energy consumption [10, 19, 43]
- Battery-related cost [27, 32]
- Electricity cost [10, 27]

Table 1 Cost per kilometer comparison of different sources[4]

Source	Cost per kilometer (EUR)	Comment
MacKenzie et al. [29]	0.36	General AV cost compared to manual vehicle
Spieser et al. [39]	0.25	SAV compared to PAV and manual vehicle
Chen et al. [9]	0.23–0.27	SAV in different charging scenarios
Stephens et al. [38]	0.10–0.17	SAV in different cost scenarios
Bösch et al. [6]	0.07–0.11	Cost range regarding pooled and private SAV
Fagnant and Kockelman [20]	0.28	No differentiation of cost parameters

- Maintenance and repair [6, 9, 27]
- Cleaning [6, 27]
- Insurance [27, 43]
- Parking [5, 11].

Taking these parameters into account, studies have found cost per kilometer to be as low as 0.07 EUR in a pooled SAV scenario to 0.36 EUR in a private AV scenario. Table 1 shows an overview of sources with different approaches and foci. However, all presented sources indicate that SAVs have a significant cost-reduction potential per kilometer, as costs are distributed among several passengers.

Presumably owing to insufficient data, none of the studies presented have incorporated actual operational costs of manual or AV ride-hailing fleet operators, and mostly rely on estimates. Furthermore, most of the studies focus on cities or regions in the USA, and transferability to European or German markets may be limited. Hence, application of their findings to urban areas in Germany should be treated cautiously.

2.4 Profitability Analyses of Shared Autonomous Vehicle Services

While studies have shown the potential cost-effectiveness of AV technology for fleet operators, operators will only enter the market if they see a profitable business, at least in the long-term. Cost-effectiveness does not necessarily mean that market conditions remain constant. To the authors' knowledge, Chen et al. [9], Bösch et al. [6] and Loeb and Kockelman [27] are the only studies to date to analyze the potential

---

[4] The findings were converted from the original sources with 1 USD = 0.90 EUR (as of 1st May 2020) and 1 mile = 1.61 km.

profitability of SAV services. [39] emphasized the need of operators to eventually run a profitable fleet, but focused on cost implications of AVs and neglected market developments. Furthermore, they did not include cost for fleet operations. As a valuable reference for fleet demand, they modeled up to 300,000 AVs for the case of Singapore, assuming a scenario of 100% market penetration of AVs [39]. Analogously, Chen et al. [9] analyzed an abstract area of 10,000 square miles (~26,000 km^2). Based on a 10% market share, their model estimated that more than 57,000 vehicles would be needed. For profitability analysis, they assumed a constant 10% operating margin, which resulted in customer prices per occupied mile of 0.66–0.74 USD (~0.37–0.42 EUR/km). However, constant operating margins imply that operators have the indefinite market power to set increasing customer prices corresponding to increasing cost structures. Loeb and Kockelman [27] recognized this limitation, and assumed customer prices of 1 USD/mile and derived profits per revenue-mile of 0.60 USD in a low-cost and 0.14 USD in a high-cost scenario.

Similar to the research on cost structures of SAVs, studies have almost exclusively focused on cities in the USA, and usually use the same models and cities to analyze the impact of AVs in mobility systems from different perspectives. The studies note that application of their models to other cities should be limited to regions with similar demographics, travel, and land-use patterns [27]. Differences in city characteristics, mobility behavior and attitudes of customers and policymakers in European cities may lead to considerably different business implications for commercial AV fleet operators.

3   Data

Market analysis

For our market analysis, fleet and trip data provided by operational carsharing providers and ride-hailing providers—including conventional taxi businesses—as well as their respective German federal associations (Bundesverband Taxi und Mietwagen e.V., Bundesverband CarSharing e.V.) was used. The trip data, which the carsharing and ride-hailing vehicles automatically and anonymously transmit via their telematic systems or taxi meters, was retrieved from a sample of 2,100 ride-hailing vehicles and 3,385 carsharing vehicles. The sample data was recorded and shared with us for the full year 2019 in different cities in Germany. To estimate the overall market demand for Germany, average values of the sample for the number of trips per vehicle, respective trip lengths and occupancy rates, as well as utilization and travel speed were calculated and applied to the total German fleet. The data was further complemented by and triangulated with publicly available business reports, and interviews with industry experts and executives of the mobility providers. To estimate the relative share of shared vehicle services, data from the MiD 2017 was used, which is the latest study on household mobility in Germany. For the scenario setup, we used UN forecasts for urban population development until 2035.

Supply model

For the supply model, we additionally used actual weekly demand and speed patterns on an hourly basis, which was again recorded and shared with us by operational carsharing and ride-hailing providers. The data was further validated with findings published by Uber and Lyft. Since both TNCs went public in 2019, this study benefits from the increasing amount of the publicly available data. Average trip length and passenger occupancy rates were also derived from the sample data described above.

Cost structure analysis

The cost structure analysis was based on quarterly and yearly financial reports and operational Key Performance Indicators (KPI) reports of leading mobility providers in Germany. Detailed damage reports and cleaning procedure documents were also shared with us where the financial reports did not allow for a detailed analysis. Reports from 2017 to 2019 were provided and analyzed. To complement and triangulate cost parameters, such as insurance costs, subject matter experts were interviewed and asked for their assessment of the respective cost parameter. Furthermore, publicly available data of the TNCs Uber and Lyft were used. For the projection of every cost parameter (n = 41), 31 subject matter experts were interviewed individually. Their projections were further compared to statements by executives of market participants, such as Waymo, AutoX, Deeproute.ai, Baidu or Mobileye. The statements were publicly available on respective company blogs and YouTube accounts, press releases or newspaper interviews.

4 Methodology

For the assessment of the current market demand for shared vehicle services in Germany, free-floating and station-based carsharing and ride-hailing services (including traditional taxi services) were considered. To allow for efficient analysis of the sample (see data section), the data was filtered for urban areas with at least 100,000 inhabitants (n = 80). In terms of total fleet size, these areas account for 99% of all vehicles. The full year 2019 was selected as the evaluation period as it presents the latest year with comparable data quality for all considered service types. To estimate the overall market demand for Germany, average values of the sample were calculated and applied to the total fleet size estimates provided. While the carsharing demand was derived based on the average utilization[5] and travel speed of the sample, ride-hailing demand was estimated based on the number of trips per vehicle, respective trip lengths and occupancy rates of the sample.

Due to the high level of uncertainty in the future development of shared vehicle services, a basic scenario approach was taken to examine different future market demands for SAV services in Germany. Based on the results of the current market

---

[5] Utilization in this paper refers to the vehicle time with a customer on board, based on a 24-h day.

demand analysis, three scenarios were derived, which describe different developments for the demand of SAV services for the year 2035. The share of SAV services in the total of all passenger kilometers (modal split) in the urban area of Germany was used as the elementary parameter for the differentiation of the scenarios:

1. Neutral: SAV service share remains at a constant level of 0.5% in 2035
2. Evolution: SAV service share increases to 2.5% in 2035
3. Revolution: SAV service share increases to 10% in 2035.

The Neutral scenario serves as the base scenario. Compared to the base case, the demand for SAV services is five times higher in the Evolution scenario and 20 times higher in the Revolution scenario. Based on United Nations (UN) forecasts for urban population development, a population increase of 2% is assumed for all three scenarios for the simulated period until 2035. We assume that this will result in an increase in urban demand for mobility of the same magnitude. While the overall German population is expected to decrease until 2035 by around 1 million people, the urbanization trend is expected to continue, according to the UN.

The different projected SAV service shares of the modal split serve as input variables for the supply model to examine how a potential SAV fleet operator can reasonably serve future demand. The supply model to retrieve crucial fleet operation data, such as required fleet sizes, vehicle mileage and utilization rates, is established using the System Dynamics (SD) method. SD was selected as the method allows for describing, modeling, simulating, and analyzing dynamically complex issues such as the operation of an SAV service over time [33]. One crucial advantage of the SD method is the option to implement feedback loops in addition to unidirectional causalities. Hence, the behavior of variables of the system can be (partly) caused by its own behavior in the past. This may lead to non-linear behavior in the overall system, even if all constitutive causalities are linear. To allow for comparability of the different market share scenarios, the following input parameters were held constant:

- Weekly demand patterns (hourly basis, weighted average of data sample)
- Weekly speed patterns (hourly basis, weighted average of data sample)
- Average trip length (7.6 km, weighted average of data sample)
- Maximum waiting time for passengers (5 min)
- Average passenger occupancy rate (1.5, weighted average of data sample).

The established model is simulated over time using SD simulation software. In this study, the software PowerSim was used for simulation.

In respect of the increasing electrification trend of the total car parc in Germany, maintenance advantages of battery-electric vehicles (BEV) in the urban fleet context, and reflecting the increasing regulatory pressure to reduce CO_2 emissions, this study assumes that all SAVs of the service's fleet are BEVs.

To derive the corresponding cost structure for the provider's fleet operation, actual operational data from shared vehicle services and pilots was used as a basis, including data of free-floating and station-based carsharing and ride-hailing in Germany. In cases of missing or greatly varying parameters, plausible assumptions were made.

Costs for research and development as well as overhead costs were not further considered as they vary greatly and may present considerable administrative inefficiencies. All derived cost parameters (n = 41) were further triangulated by subject matter experts and assessed for their applicability in an SAV fleet. Furthermore, their development until 2035 was estimated by subject matter experts in individual interviews and group discussions to yield a multi-perspective projection.

5 Results

5.1 Current Demand for Shared Vehicle Services in Germany and Three Scenarios for 2035

According to our sample assessment and respective extrapolation, the current demand for shared vehicle services in Germany is around 3.6 billion PKM per year. This includes the demand for free-floating and station-based carsharing and ride-hailing. Based on the most current available data from the 2017 Mobility in Germany study ("Mobilität in Deutschland", MiD), this results in a relative share of 0.5% of the total yearly transport performance of the urban population in Germany.[6] Those values represent an average for the urban area of Germany overall, without spatial differentiation. As shared vehicle services are sometimes only offered in inner-city business areas (e.g., free-floating carsharing), the distribution in different city areas can vary greatly. The absolute demand was served by around 119,000 shared vehicles. The respective total yearly revenue is estimated to be around 5.7 billion EUR.

The current demand for shared vehicle services in Germany of 3.6 billion PKM also served as the basis for the subsequent scenario derivation. An overview of the derived market demand in the three scenarios can be found in Fig. 2. In all three scenarios, the mobility demand for all modes in 2035 is assumed to be 2% higher compared to 2019, as the urban population in Germany is expected to grow at the same magnitude. Hence, although the market share in the Neutral scenario remains at a constant level, it still results in a higher market demand of 3.8 billion PKM.

5.2 Supply Model Results: Operational Implications for an SAV Service Provider to Meet the Demand in 2035

The supply simulation model examines how an SAV service provider would reasonably serve the derived market demand of the three scenarios in 2035. Hence, the different market demands constituted the starting points for the supply simulation,

---

[6] The total yearly transport performance of the urban population in Germany is 742.4 billion passenger kilometers (PKM).

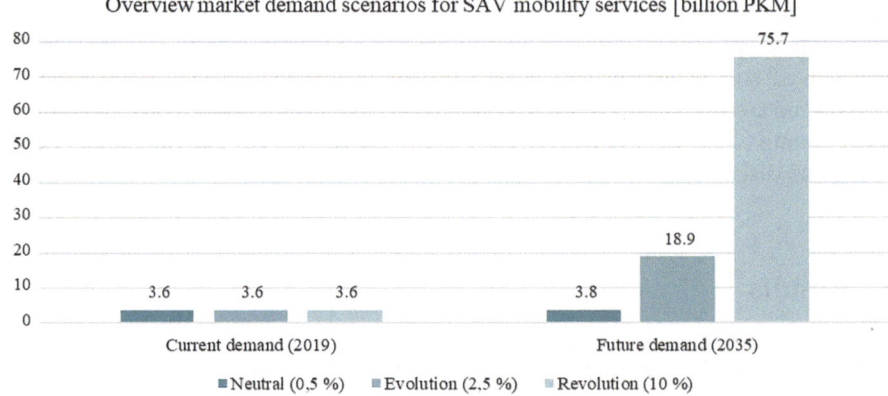

Fig. 2 Overview of market demand scenarios in billion PKM for SAV services in Germany

which was run multiple times for each of the three scenarios. The findings are presented in Table 2. To balance required fleet sizes with the maximum waiting time for passengers of 5 min, the service level was set to 96%—meaning that 4% of the market demand was not served.

The weekly demand patterns peak in the mornings and afternoons, resulting in a vehicle utilization ranging from 25% in the Neutral scenario to 32% in the Revolution scenario. The higher utilization rate is mainly due to the higher demand density within the considered area. The required fleet size in the Neutral scenario of 40,944 vehicles shows that only around one-third of the current fleet of 119,000 shared vehicles would be needed to serve the same demand if the SAV diffusion rate was 100%. Hence, the car parc could be reduced by more than 78,000 vehicles without compromising customer transport needs. In the Revolution scenario, with 10% of the total urban transport performance in Germany being attributed to SAVs, a fleet of 639,749 vehicles would be needed.

Table 2 Overview of selected output parameters of the supply model

	Neutral scenario	Evolution scenario	Revolution scenario
Market demand (PKM)	3,786,291,000	18,931,455,000	75,725,820,000
Service level	96%	96%	96%
Served market demand (PKM)	3,634,839,360	18,174,196,800	72,696,787,200
Served trips (#)	318,845,558	1,594,227,789	6,376,911,158
Per vehicle kilometres in service (VKM)	59,184	68,653	75,756
Ø Vehicle utilization	25%	29%	32%
SAV fleet size needed (#)	40,944	176,483	639,749

5.3 Financial Analysis: Derivation of Respective Cost Per Vehicle-Kilometer and Profitable Price Levels in 2035

To ensure comparability with other studies, the cost of the SAV service operation was first calculated on a per VKM basis. The output parameters of the supply simulation model for each scenario in 2035 served as reference values. Yearly, monthly, and daily per vehicle costs were multiplied with the fleet sizes, accumulated for the full year and divided by the total mileage in 2035. Similarly, trip-based costs were accumulated with the number of yearly trips served and standardized to the yearly mileage. Finally, cost parameters on per VKM basis were added. Hence, the results show the cost per VKM for the full business year 2035. As for the vehicle costs, a monthly leasing rate was calculated, assuming a maximum lifetime mileage of 350,000 km, an interest rate of 2.5% and a dismantling residual value of 5%.

Table 3 shows the total cost per VKM and the shares of respective cost parameters. It must be noted that the cost parameter projections for 2035 reflect the optimistic assessments by subject matter experts. Hence, the results should be considered to show the lower boundary of the cost range. Moreover, the costs do not include administrative fixed costs, or expenses for research and development.

As a result, the SAV fleet may be operated at 0.48 EUR in the Revolution scenario, at 0.49 EUR in the Evolution scenario and at 0.50 EUR in the Neutral scenario. Cost for the mobility service provision (including variable customer app cost and variable marketing cost), vehicle cost and cost for electricity account for the highest cost shares.

To derive profitable price levels for all scenarios in 2035, the same cost basis was used. However, the cost per VKM was derived using only VKMs in service, i.e., with paying passengers on board. The calculation was conducted using the cost base for the full year 2035 divided by the sum of VKMs when the SAV fleet was occupied in the same year. This approach takes the average occupancy rate of 1.5 passengers per

Table 3 Cost per vehicle-kilometer in 2035 per scenario

	Neutral scenario	Evolution scenario	Revolution scenario
Cost per VKM [EUR]	0.50	0.49	0.48
Thereof mobility service provision	29%	30%	30%
Thereof vehicle cost	28%	28%	28%
Thereof insurance cost	3%	3%	3%
Thereof maintenance and repair cost	11%	10%	10%
Thereof cleaning cost	4%	4%	3%
Thereof electricity cost	17%	17%	18%
Thereof IT operation cost	3%	3%	3%
Thereof staging and teleoperation cost	4%	4%	3%

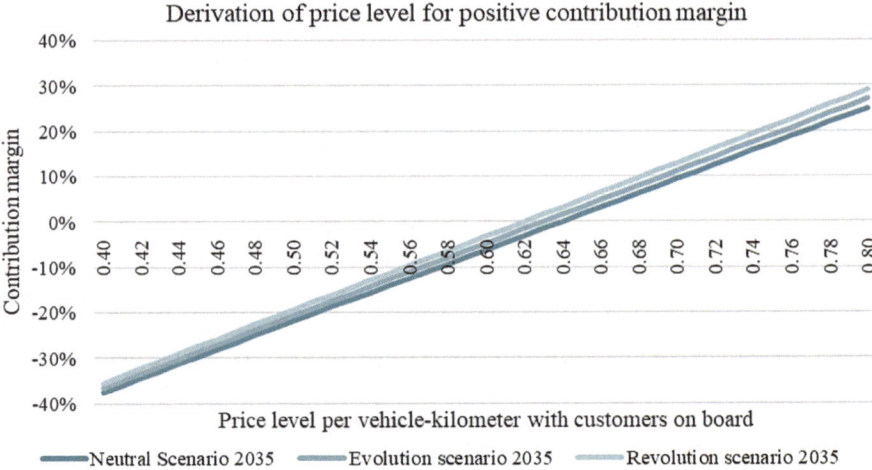

Fig. 3 Derivation of price levels for positive contribution margins per scenario

trip into account, to enable a certain degree of comparability with current ride-hailing offers which take a similar pricing strategy. However, in contrast to traditional taxi services, fixed fees were neglected. Moreover, as overheads are not considered in this analysis, the term profitability refers to positive contribution margin levels.

As shown in Fig. 3, contribution margins turn positive at customer prices of 0.62 EUR in the Revolution scenario, 0.63 EUR in the Evolution scenario and 0.64 EUR in the Neutral scenario.

6 Discussion of Results and Limitations

Compared against the urban transport performance shares of ride-hailing providers in the USA, which range from 2.6% in Los Angeles to 19% in New York City, the 0.5% share of shared vehicle services in Germany is considerably lower, and is expected to remain at a lower level—even with the introduction of SAVs [3, 36]. Nevertheless, the Evolution and Revolution scenarios are intended to illustrate possible future demands, driven by high transport comfort, easy and on-demand accessibility, and cities working hard to relieve the urban transport system by banning motorized individual transport. From a neutral point of view, we leave it up to the corporate strategists to evaluate the probability of occurrence of the future scenarios. The aim here was to support the strategy process by deriving quantitative operational and cost implications for future demand scenarios.

The results of the supply model may also help urban policymakers in their evaluation of SAV services, as the results show significant efficiency gains with SAVs in the fleet. Only around one-third of the current fleet of 119,000 shared vehicles would be needed to serve today's demand if the SAV diffusion rate was 100%, due to higher utilization rates of the vehicles. Although current ride-hailing providers achieve similar utilization rates at a shift level (the data sample showed an average shift time of 8.3 h), vehicles remain unproductive off-shift. This leads to a significantly lower utilization over the course of 24 h. The comparison of scenario results shows that the utilization of SAVs increases even more when the demand density in the urban area grows: while the market demand of the Revolution scenario is 20 times higher compared to the Neutral scenario, the required fleet size to serve the demand is only 15.6 times higher. SAV fleet operators may further increase the utilization rate by lowering their service level, especially during the peak times of the daily demand patterns ('peak shaving'). However, rejection of passenger requests would most certainly result in a bad customer experience. Thus, fleet operators are required to constantly balance operational efficiency and customer experience.

The financial analysis reveals that according to our data, SAV fleets may be operated at per VKM cost of 0.48–0.50 EUR. The differences across scenarios are again mainly due to the different utilization rates of the vehicles. In comparison to other literature results, the cost range exceeds all the findings presented by MacKenzie et al. [29], Spieser et al. [39], Chen et al. [9], Stephens et al. [38], Bösch et al. [6] and Fagnant and Kockelman [20]. Compared to the highest cost finding, 0.36 EUR/VKM stated by MacKenzie et al. [29], our results suggest 33–39% higher costs. This is mainly caused by the fact that we take into account the cost of mobility service provision, such as running a customer app or variable marketing cost per trip. Neglecting those costs would result in a cost range of 0.34–0.36 EUR, which is in line with the upper cost boundary of the literature findings. It must be stated, however, that our cost predictions for 2035 are based on the optimistic projections of the interviewed subject matter experts. From an OEM (Original Equipment Manufacturer) perspective, the number of new vehicles sold per year is an important parameter to evaluate the market demand at vehicle level, and whether or not investment in developing a suitable vehicle is cost-effective. While fleet sizes indicate a volume ranging from 40,944 SAVs in the Neutral scenario to 639,479 SAVs in the Revolution scenario, the number of new SAVs needed per year amounts to 8,931 and 178,626 in 2035. Based on the average yearly number of new car registrations in Germany from 2017–2021, this results in a sales volume share of 0.3% and 5.6%, respectively.

Furthermore, the financial analysis shows that customer prices of 0.62 EUR in the Revolution scenario, 0.63 EUR in the Evolution scenario and 0.64 EUR in the Neutral scenario are necessary to reach the threshold of positive contribution margins. These price levels per occupied VKM present a significant reduction as compared to the average price level of 2.35 EUR of the ride-hailing data sample. If this price level is held constant until 2035, it suggests attractive absolute contribution margins per VKM of 1.71–1.73 EUR, depending on the scenario. However, overhead and development costs may consume a major share of the margin, especially if a purpose-built vehicle is to be developed. Furthermore, the constant price level may only be achieved in a monopoly market. Assuming several market participants compete for market shares, the derived cost advantage may considerably decrease over time. Hence, OEMs, AV technology providers and fleet operators need to continue their efforts to identify further cost saving potential. OEMs may directly influence two of the biggest operational cost drivers by working on durable vehicle designs and the efficiency of the electric powertrain.

There are various limitations to our approach. Firstly, the market analysis was conducted based on data from 2019. Thus, it relies on data before the COVID-19 pandemic, which does not reflect the impact of the crisis on mobility behavior. As we do not have sufficient quantitative data on this matter, we asked executives of the mobility market for a qualitative assessment. According to those interviews, the COVID-19 crisis has affected shared vehicle services due to hygiene concerns (sharing the ride or the vehicle with others) and less mobility in general. Interestingly, the number of carsharing vehicles is still rising but on a lower level, presumably due to substitution effects from public transport to more privately shared options. The MiD data as well as the UN urbanization projections may also be affected by the COVID-19 pandemic, as they date back to 2017 and 2019, respectively. Moreover, in our straightforward scenario derivation, we do not consider psychological parameters as to how much passengers trust AVs. However, this may certainly play a role when deciding between different transport modes (see also chapter "Social Acceptance of CAD in Japan and Germany: Conceptual Issues and Empirical Insights"). As for the system dynamics supply model, it does not allow for spatial differentiation within the urban areas (e.g., different demand for SAVs in different districts). The study area is rather regarded as one aggregated average area. Finally, although the cost structure analysis was done using actual operational data, and projections were based on interviews with subject matter experts for the different cost parameters, a certain degree of uncertainty remains. There may also be additional cost parameters which are not yet visible.

7 Conclusion

In the present study, we used a detailed operational dataset of carsharing and ride-hailing vehicles to analyze the current shared vehicle service market in Germany. Using the current market results as a basis, we set up simplified scenarios to derive possible market demands in 2035, and used a system dynamics (SD) simulation approach to analyze the operational implications of serving those future demands with shared AVs. Eventually, we applied a cost structure analysis and prognosis to assess price levels at which an SAV fleet operation is profitable. While previous studies mainly focused on modeling and simulating mobility markets in the USA, our goal was to shift this focus to a European country with well-established public transport and a regulated market. Moreover, benefiting from the maturing mobility market in Germany, we added actual operational data to the existing literature on the cost structure of SAV fleet operations.

The results indicate that the current market of shared vehicle services accounts for 0.5% of the total urban transport performance in Germany, which is considerably lower compared to urban areas in the USA. In absolute terms, carsharing and ride-hailing services accumulate around 3.6 billion PKM. Serving market demands in

2035 ranging from 3.8–75.7 billion PKM across different scenarios, a fleet of 40,944–639,749 SAVs would be required. Due to higher utilization rates of the SAVs, only around one-third of the current fleet of 119,000 shared vehicles would be needed to serve today's demand if the SAV diffusion rate was 100%. We also found that cost structures of SAV fleet operations may be highly underestimated in previous studies, resulting in a per VKM cost range of 0.48–0.50 EUR. Moreover, we derived customer price levels at which the SAV fleets are operated profitably. While these are significantly lower than the current cost of using ride-hailing services, they may not be low enough to convince a considerable number of passengers to switch from their private cars to SAV services. Regarding these results, we recommend that OEMs, AV technology providers and fleet operators continue their efforts to realize further cost savings and fleet efficiencies. However, the support of policymakers in urban areas is also needed to incentivize more people to switch from private car transport to SAV services.

As for future research, we recommend that academia should not only focus on the current business logic of passengers paying for their trip, but also on new business model approaches in and around SAVs and their respective trips. This may show additional revenue streams and secure future business success for SAV services. Furthermore, as development costs for SAVs are high, established OEMs still struggle to find their role within the mobility value chain. To overcome the dilemma caused by high development costs, mobility services which may use existing vehicle architectures should be further evaluated, such as intercity SAV services. Eventually, the SAV market is gaining further momentum due to the advancements in autonomous driving technology and an increasing number of permits to operate AVs on public roads. Hence, the cost parameters should be constantly reviewed and complemented with actual operational data.

References

1. Ammann, D. (2020, April 23). The Cruise Origin Story. *Cruise*. https://medium.com/cruise/the-cruise-origin-story-b6e9ad4b47e5.
2. Anderson, J. M., Kalra, N., Stanley, K. D., Sorensen, P., Samaras, C., & Oluwatola, T. A. (2016). *Autonomous Vehicle Technology: A Guide for Policymakers*. RAND Corporation. https://www.rand.org/pubs/research_reports/RR443-2.html.
3. Balding, M., Whinery, T., Leshner, E., & Womeldorff, E. (2019). *Estimated Percent of Total Driving by Lyft and Uber In Six Major US Regions* (Nr. SF19–1016). https://www.urbanismnext.org/resources/estimated-percent-of-total-driving-by-lyft-and-uber.
4. Bansal, P., Kockelman, K. M., & Singh, A. (2016). Assessing public opinions of and interest in new vehicle technologies: An Austin perspective. *Transportation Research Part C: Emerging Technologies, 67*, 1–14. https://doi.org/10.1016/j.trc.2016.01.019.
5. Bischoff, J., Maciejewski, M., Schlenther, T., & Nagel, K. (2019). Autonomous Vehicles and their Impact on Parking Search. *IEEE Intelligent Transportation Systems Magazine, 11*(4), 19–27. https://doi.org/10.1109/MITS.2018.2876566.

6. Bösch, P. M., Becker, F., Becker, H., & Axhausen, K. W. (2018). Cost-based analysis of autonomous mobility services. *Transport Policy, 64*, 76–91. https://doi.org/10.1016/j.tranpol. 2017.09.005.
7. Bundesministerium für Digitales und Verkehr. (2021, Juli 27). *Gesetz zum autonomen Fahren tritt in Kraft.* https://www.bmvi.de/SharedDocs/DE/Artikel/DG/gesetz-zum-autonomen-fahren.html.
8. Bunghez, L. (2015). The Future of Transportation-Autonomous Vehicles. *International Journal of Economic Practices and Theories, 5*(5), 447–454.
9. Chen, T. D., Kockelman, K. M., & Hanna, J. P. (2016). Operations of a shared, autonomous, electric vehicle fleet: Implications of vehicle & charging infrastructure decisions. *Transportation Research Part A: Policy and Practice, 94*, 243–254. https://doi.org/10.1016/j.tra.2016. 08.020.
10. Chen, Y., Gonder, J., Young, S., & Wood, E. (2019). Quantifying autonomous vehicles national fuel consumption impacts: A data-rich approach. *Transportation Research Part A: Policy and Practice, 122*, 134–145. https://doi.org/10.1016/j.tra.2017.10.012.
11. Clements, L. M., & Kockelman, K. M. (2017). Economic Effects of Automated Vehicles. *Transportation Research Record, 2606*(1), 106–114. https://doi.org/10.3141/2606-14.
12. Clewlow, R. R., & Mishra, G. S. (2017). *Disruptive Transportation: The Adoption, Utilization, and Impacts of Ride-Hailing in the United States.* https://escholarship.org/uc/item/82w2z91j.
13. Corwin, S., Jameson, N., Pankratz, D. M., & Willigmann, P. (2016). *The future of mobility: What's next?* (Deloitte University Press). https://www2.deloitte.com/content/dam/insights/us/articles/3367_Future-of-mobility-whats-next/DUP_Future-of-mobility-whats-next.pdf.
14. Deeproute.ai. (2020, August 19). *DeepRoute links up with Cao Cao Mobility to roll out robotaxi fleet in Hangzhou* [Official Company Website]. https://deeproute.ai/en/news/detail.php?id=4&bid=41&cid=250.
15. Deloitte. (2019). *Urbane Mobilität und autonomes Fahren im Jahr 2035* [Studie Unternehmensberatung]. https://www2.deloitte.com/content/dam/Deloitte/de/Documents/Innovation/Datenl and%20Deutschland%20-%20Autonomes%20Fahren_Safe.pdf.
16. Deloitte (Hrsg.). (2021). *Nachfrage sucht Angebot Pragmatismus beim Aufbau von Mobilitätsökosystemen.* https://www2.deloitte.com/content/dam/Deloitte/de/Documents/consumer-industrial-products/Mobilit%C3%A4ts%C3%B6kosysteme-Nachfrage-sucht-Angebot.pdf.
17. Enoch, M. P. (2015). How a rapid modal convergence into a universal automated taxi service could be the future for local passenger transport. *Technology Analysis & Strategic Management, 27*(8), 910–924. https://doi.org/10.1080/09537325.2015.1024646.
18. Fairfield, N. (2016, Dezember 13). *On the road with self-driving car user number one.* Waypoint-The official Waymo blog. https://blog.waymo.com/2016/12/on-road-with-self-driving-car-user.html.
19. Fagnant, D. J., & Kockelman, K. (2015). Preparing a nation for autonomous vehicles: Opportunities, barriers and policy recommendations. *Transportation Research Part A: Policy and Practice, 77*, 167–181. https://doi.org/10.1016/j.tra.2015.04.003.
20. Fagnant, D. J., & Kockelman, K. M. (2018). Dynamic ride-sharing and fleet sizing for a system of shared autonomous vehicles in Austin, Texas. *Transportation, 45*(1), 143–158. https://doi.org/10.1007/s11116-016-9729-z.
21. Fox, S. (2016). Planning for Density in a Driverless World. *IPR Papers & Reports.* https://scholarship.law.georgetown.edu/ipr_papers/1.
22. Handelsblatt. (2021, September 7). *Mobileye und Sixt bringen Robotaxis 2022 nach Deutschland.* https://www.handelsblatt.com/unternehmen/industrie/iaa-mobileye-und-sixt-bringen-robotaxis-2022-nach-deutschland/27590814.html.
23. Heineke, K., Heuss, R., Kampshoff, P., Kelkar, A., & Kellner, M. (2022). *The road to affordable autonomous mobility* (S. 9).
24. Jones, E. C., & Leibowicz, B. D. (2019). Contributions of shared autonomous vehicles to climate change mitigation. *Transportation Research Part D: Transport and Environment, 72*, 279–298. https://doi.org/10.1016/j.trd.2019.05.005.

25. Kaas, H.-W., Mohr, D., Gao, P., Müller, N., Wee, D., Hensley, R., Guan, M., Möller, T., Eckhard, G., Bray, G., Beiker, S., Brotschi, A., & Kohler, D. (2016). *Automotive revolution–perspective towards 2030* (Advanced Industries). https://www.mckinsey.com/~/media/mck insey/industries/automotive%20and%20assembly/our%20insights/disruptive%20trends%20t hat%20will%20transform%20the%20auto%20industry/auto%202030%20report%20jan% 202016.pdf.
26. Lang, N., Herrmann, A., Hagenmaier, M., & Richter, M. (2020). *Can Self-Driving Cars Stop the Urban Mobility Meltdown.* https://image-src.bcg.com/Images/BCG-Can-Self-Driving-Cars-Stop-the-Urban-Mobility-Meltdown-Jul-2020_tcm74-252485.pdf.
27. Loeb, B., & Kockelman, K. M. (2019). Fleet performance and cost evaluation of a shared autonomous electric vehicle (SAEV) fleet: A case study for Austin, Texas. *Transportation Research Part A: Policy and Practice, 121*, 374–385. https://doi.org/10.1016/j.tra.2019.01.025.
28. Machado, C. A. S., De Salles Hue, N. P. M., Berssaneti, F. T., & Quintanilha, J. A. (2018). An Overview of Shared Mobility. *Sustainability, 10*(12), 4342. https://doi.org/10.3390/su1012 4342.
29. MacKenzie, D., Wadud, Z. & Leiby, P. N. (2014). A first order estimate of energy impacts of automated vehicles in the United States. In: Transportation research board annual meeting, Vol. 93rd. The National Academies, Washington, DC.
30. McKinsey Center for Future Mobility (Hrsg.). (2019). *The future of mobility is at our doorstep.* https://www.mckinsey.com/~/media/McKinsey/Industries/Automotive%20and% 20Assembly/Our%20Insights/The%20future%20of%20mobility%20is%20at%20our%20d oorstep/The-future-of-mobility-is-at-our-doorstep.ashx.
31. Metropolitan Area Planning Council of the City of Boston (MAPC) (Hrsg.). (2018). *Share of Choices—Further Evidence of the ride-hailing effect in Metro Boston and Massachusetts.* https://www.mapc.org/planning101/shareofchoices/.
32. Nykvist, B., & Nilsson, M. (2015). Rapidly falling costs of battery packs for electric vehicles. *Nature Climate Change, 5*(4), 329–332. https://doi.org/10.1038/nclimate2564.
33. Pruyt, E. (2013). *Small System Dynamics Models for Big Issues: Triple Jump towards Real-World Dynamic Complexity* (1. Aufl.). TU Delft Library.
34. Richter, M. A. (2020). Autonome Fahrzeuge als Lösung für die heutige Verkehrsbelastung. *Internationales Verkehrswesen, 72*(4), Article 4. https://www.alexandria.unisg.ch/262195/.
35. Rode, P. (2013). Trends and Challenges: Global Urbanisation and Urban Mobility. In *Megacity Mobility Culture: How Cities Move on in a Diverse World* (S. 3–21). Springer. https://doi.org/ 10.1007/978-3-642-34735-1_1.
36. Schaller, B. (2017). *Unsustainable? The Growth of App-Based Ride Services and Traffic* (Travel and the Future of New York City). https://www.urbanismnext.org/resources/unsustainable-the-growth-of-app-based-ride-services-and-traffic-travel-and-the-future-of-new-york-city.
37. Soteropoulos, A., Berger, M., & Ciari, F. (2019). Impacts of automated vehicles on travel behaviour and land use: An international review of modelling studies. *Transport Reviews, 39*(1), 29–49. https://doi.org/10.1080/01441647.2018.1523253.
38. Stephens, T. (2016), Estimated Bounds and Important Factors for Fuel Use and Consumer Costs of Connected and Automated Vehicles, Technical Report, National Renewable Energy Laboratory.
39. Spieser, K., Treleaven, K., Zhang, R., Frazzoli, E., Morton, D., & Pavone, M. (2014). Toward a Systematic Approach to the Design and Evaluation of Automated Mobility-on-Demand Systems: A Case Study in Singapore. In G. Meyer & S. Beiker (Hrsg.), *Road Vehicle Automation* (S. 229–245). Springer International Publishing. https://doi.org/10.1007/978-3-319-05990-7_ 20.
40. Templeton, B. (2022, Januar 25). *Robocars.com interview with Amnon Shashua of MobilEye on 4 key issues.* https://www.youtube.com/watch?v=UfkxNFHG0ys.
41. Trommer, S., Kolarova, V., Fraedrich, E., Kröger, L., Kickhöfer, B., Kuhnimhof, T., Lenz, B., & Phleps, P. (2016). *Autonomous Driving—The Impact of Vehicle Automation on Mobility Behaviour* [Berichtsreihe]. http://www.ifmo.de/publications.html?t=45.

42. United Nations. (2019). *World urbanization prospects: The 2018 revision.* Department of Economic and Social Affairs, Population Division.
43. Wadud, Z. (2017). Fully automated vehicles: A cost of ownership analysis to inform early adoption. *Transportation Research Part A: Policy and Practice, 101,* 163–176. https://doi.org/10.1016/j.tra.2017.05.005.

Social Acceptance of CAD in Japan and Germany: Conceptual Issues and Empirical Insights

Torsten Fleischer, Ayako Taniguchi, Jens Schippl, Yukari Yamasaki, Kosuke Tanaka, and Satoshi Nakao

Abstract It is widely acknowledged that social acceptance of automated vehicles (AVs) is a crucial factor for the future development and deployment of the technology in mobility systems. In general, mobility systems are sociotechnical systems. Their design and development depend on a multitude of technical and non-technical factors, including aspects of public or social acceptance. However, as will be shown in this chapter, social acceptance can have different meanings and can be addressed by various approaches. Different objects of acceptance (e.g. trust in robots, AVs as a useful means of transport etc.) as well as different subjects of acceptance (users, citizens, industrial interest groups etc.) can be distinguished. In addition, the subjects can be in different relationships to the objects (use, approval, protest etc.). Against this backdrop, we start this chapter with an in-depth conceptualization of social acceptance. Following this, we present empirical material that sheds lights on different dimensions of acceptance. We draw on two surveys carried out in Japan and Germany in recent years. The surveys provide insights on relevant public perceptions and attitudes towards AVs, and make it clear that not only public perceptions but also the views and attitudes of many other actors are relevant for acceptance and diffusion of

T. Fleischer (✉) · J. Schippl · Y. Yamasaki
Karlsruhe Institute of Technology, Karlsruhe, Germany
e-mail: torsten.fleischer@kit.edu

J. Schippl
e-mail: jens.schippl@kit.edu

Y. Yamasaki
e-mail: yukari.yamasaki@kit.edu

A. Taniguchi
University of Tsukuba, Tsukuba, Japan
e-mail: taniguchi@risk.tsukuba.ac.jp

K. Tanaka · S. Nakao
Kyoto University, Kyoto, Japan
e-mail: tanaka.kosuke.6k@kyoto-u.ac.jp

S. Nakao
e-mail: nakao@trans.kuciv.kyoto-u.ac.jp

© The Author(s) 2025
C. Eisenmann et al. (eds.), *Acceptance and Diffusion of Connected and Automated Driving in Japan and Germany*, https://doi.org/10.1007/978-3-031-59876-0_5

AVs. To give an additional perspective on this topic, we provide insights on media reporting on AVs in Japan and Germany.

1 Conceptual Issues of Social Acceptance

If automated vehicles (AVs) are increasingly used in the next few years, then citizens will encounter them in different contexts, and whilst in different roles, for example as passengers of these vehicles, as non-users participating in road traffic (drivers, cyclists or pedestrians), as buyers, or as citizens with demands on the use of public space or different varying safety expectations. Acceptance issues can become relevant anywhere, since the mobility system is a sociotechnical system in which almost every citizen participates, and everyone is affected by changes to it.

Nevertheless, from the perspective of innovation and innovation policy research, the recurring recourse to the "social acceptance" of automated driving (AD) (or, as it called in some jurisdictions, connected and automated driving (CAD)) in public statements by business leaders or politicians, in consulting studies and policy papers is surprising. For example, in 2017 the German Federal Ministry of Transport argued in its report on the implementation status of the strategy for CAD that "*social dialogue and the creation of acceptance are central prerequisites for the successful introduction of automated and connected vehicles in public road transport*" [1]. This view is also supported in the Action Plan for Research on Automated and Connected Driving 2019: "*A systemic view of mobility reveals suitable starting points and indications of necessary framework conditions [for automated and connected driving]. It is clear that technical progress in the service of safety, sustainability and user-friendliness must not be at odds with affordability, availability and social acceptance*" [4]. In the same vein, Guy Pratt, CEO of the Toyota Research Institute, commented in 2017 that, "*Social acceptance is another challenge. Not everyone is ready to embrace AD/ AI. However, we start to see a change in mindset.*" [22]. The Strategic Headquarters for the Promotion of an Advanced Information and Telecommunications Network Society (IT Strategic Headquarters) within the Japanese Cabinet also stated "*[W]hen introducing automated driving systems into society as a new technology, it is indispensable not only to develop institutions as mentioned above but also to secure social acceptance*" [12].

What usually remains open is what exactly is meant by acceptance, or specifically by social or societal acceptance, in terms of CAD. Various authors have pointed out the multi-layered nature of the term on closer examination [15]. This also applies to its use in the sociotechnical system of mobility [3, 8], and in the context of AD [8, 9]. In the German context, a definition of "social acceptance" is rarely provided, and the definition does not seem to have been actively discussed and pursued by academia or policymakers in Japan either. A Japanese dictionary, Digital Daijisen, provides a rather simple definition, "companies, facilities, new technologies, etc. to be accepted with the understanding and approval/endorsement by a local community or the public." [26], which does not distinguish between acceptance and acceptability in either terminology or meaning.

The attention to CAD in research policy circles as well as in the media during recent years has initiated a whole series of acceptance studies, which refer to automated- or autonomous driving; despite the differences in technical implementation and potential use cases, these terms are also mostly used interchangeably in the academic literature on acceptance. The acceptance studies in turn have become the subject of numerous review publications (like Becker and Axhausen [2], Gkartzonikas and Gkritza [10], Nastjuk et al. [21]). These studies focus on different acceptance topics: Consumer acceptance, customer acceptance, end-user acceptance, public acceptance, and social acceptance are among the terms regularly used.

Our own qualitative analysis of public and academic discourses on AV/AD/CAD has shown that the term "social acceptance" is used in this context with at least three different meanings:

- as a prerequisite for the deployment or diffusion of AD technologies and services to achieve related policy goals such as fulfilling the "four societal promises" of AD (improve traffic safety, make traffic and the transportation system more efficient and reduce its environmental impact, enable (individualized) mobility even for those population groups that have been excluded from it up to now, and permit new forms of time use while moving around), or strengthening the national innovation system ("public policy perspective"),
- as a prerequisite for the successful introduction and diffusion of AD technologies and services to achieve business goals such as new products and services, profits, avoiding sunk costs, acquiring a social license to operate AV-based services, or meeting corporate social responsibility goals ("business perspective"); and
- as a metaphor for dealing with moral issues, value conflicts, and acceptability [11] in the context of AD ("ethical perspective").

Accordingly, the impressive corpus of the acceptance literature offers a variety of interesting insights in detail. At the same time, it also presents a number of areas for improvement:

- Many studies avoid presenting an explicit definition of the concept of acceptance they have used. For experts, this can often be derived from the content and goal of the work, or at least plausibly assumed, but this makes comparison between the results of different studies difficult. Especially in an interdisciplinary discussion context, this approach remains a challenge for many readers, and makes the discourse and its media interpretation susceptible to misunderstandings.
- In many acceptance studies, the objects of acceptance (see below) remain underconceptualized. Our qualitative empirical research on perceptions of and attitudes toward AD suggests that the argumentation structures of citizens may be focused on at least three different groups of acceptance objects: (a) the vehicle itself, including its safety aspects as well as its situational behaviour in traffic, (b) expectations of mobility services as part of daily life and the ascribed potential of AV to fulfil them, and (c) foundational ideas about a liveable environment and a "good

life", and the role that mobility, mobility services, and mobility technologies play in this. These three levels are closely intertwined, and their relative importance in shaping attitudes and intentions to use the technology is still unclear. For quantitative studies, however, it is of central importance to take these constellations into account when formulating questions and considering the quality and scope of the results.

- A large proportion of quantitative acceptance studies are based on convenience samples, and many were conducted in the context of vehicle demonstrations and field trials. Both of these conditions create empirical "blind spots". In many cases, convenience samples overrepresent academically educated, economically better-off, younger subjects who are known to be fundamentally more tech-savvy and less risk-averse than the general population. Surveys as part of field trials, on the other hand, target a subpopulation that has positioned itself as interested in technology simply by participating in the trials.

As stated above, there is as yet neither a definition of "social acceptance" of AD nor a general agreement on the similarities and differences between related concepts such as "public acceptance" or "user acceptance". We have therefore attempted to develop a proposal for the former. In doing so, we started from an initially rather simple idea: We use a structural approach first proposed by Doris Lucke in 1995 [15], and complement it with insights from innovation and transition research as well as from mobility studies for the case of AD.

Doris Lucke's extensive research found that most acceptance phenomena are based on a relationship (what exactly does 'accept' mean? Benefit, support, tolerate, ignore, etc.?) between acceptance subjects ('Who accepts?') and acceptance objects ('What should be accepted?'), embedded in a specific acceptance context. Behind this apparent simplicity lie more complex, interdependent structures. If one assumes, for example, that considerations and concepts of "social" acceptance are only meaningful if they go beyond the adoption perspective of the individual, then the results of diffusion research come into view [24]. From these, it is known that adoption decisions are regularly made by individuals within, and influenced by, their social networks. Therefore, it is important to better understand these actor networks and their influences on AD adoption and rejection. Qualitative empirical research has also shown that it is equally important to gain insights into the extent to which, and under what circumstances, citizens are willing to accept the use of AD by others in their immediate living environment, even if they themselves are not thinking of using it ("non-user acceptance"), and how the availability of AD could/would change their own social contexts of mobility and mobility technology decisions ("mode choice").

It is widely recognized in innovation research that the relationships and interactions between innovation actors are governed by shared habits, routines, or established practices rooted in both informal constraints (sanctions, taboos, customs, traditions, and codes of conduct) and formal rules (constitutions, laws, rights)—patterns which are summarized as institutions [25]. It has further been noted that transformative or disruptive innovations—almost by definition—require substantial redesign of existing institutional arrangements, or even the creation of new ones [17]. In the mobility field, many of these arrangements are highly stable and habitualized, and some of them are emotionally charged [27]. Externally-imposed interventions in these arrangements are in many cases perceived as unwelcome disruptions and are therefore often rejected. Against this background, the ability of networks of innovation actors to modify existing institutions, or create new ones (largely) unchallenged should also be considered as an element of social acceptance.

A similar broadening of perspective arises when considering the range of acceptance objects in the context of AD. What should be accepted? Is it a specific, clearly-defined driving function, or an AV with a range of different driving functions? Is the discussion focused on handover/acceptance strategies for partially autonomous vehicles, or on concerns about fully automated vehicles interacting with humans on the road? Is a new flexible mobility service to be based on vehicles with AD capabilities? How is the set of rules designed which will determine the behaviour of AV (and any consequences) in the event of an impending collision? Altered everyday routines due to changed mobility services and tools, or even the idea of a fully transformed, sustainable mobility system, are among the many ways in which "automated driving" is depicted, represented, or sometimes just imagined in empirical studies or public and political debates.

Our qualitative research to date shows that citizens' expectations and attitudes are often not only oriented toward the technology itself. Rather, the associated performance expectations, consequences of use, service concepts, or local "mobility futures" are usually addressed. Especially in quantitative surveys, it must be assumed that such framings implicitly influence response behaviour, but these are usually not made explicit in (or cannot be captured by) the methodology.

The considerations and findings presented above allow us to propose a working definition for further research on the social acceptance of a technology and to apply it to the scientific and social discourse on AD:

> Social acceptance of a technology can be defined as a favourable or positive response (like attitude, stated preference or action) by a given actor group or actor network (e.g. nation state, region, local community, organization), relating to a proposed or emerging technology or an imaginary of a socio-technical regime or socio-technical system modified by this technology, and the reasonable expectation to find explicit or tacit approval of the related processes of its institutionalization within specific spatial-temporal boundaries [8].

To capture the full scope of what we consider to be elements of social acceptance, consideration of acceptance subjects should be extended to professional actors (such as decision-makers in public administration and companies), whose "acceptance" in their specific roles—due to their influence on technology and system designs and procurement decisions—is also likely to be of considerable importance for successful adoption/dissemination of AD [23]. Additional attention should be paid to the role of organizations (such as public service companies, research institutes, civil society organizations (CSO), or regulators) as actors in innovation networks. Which variants of AV technologies and services they "accept" (or reject) and why they do so, will influence the acceptance heuristics of other individual or professional innovation actors.

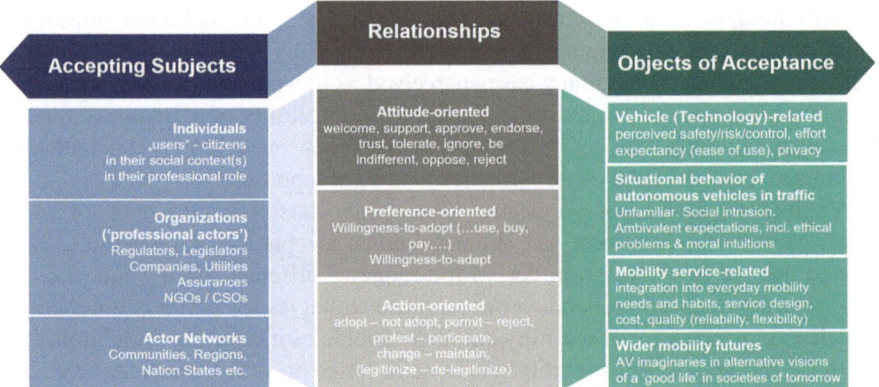

Fig. 1 Relationships between subjects and objects of acceptance in the context of automated driving

We have systematized and summarized the elements of social acceptance definition in Fig. 1. Individual research approaches cannot fully capture and illuminate
the complex fabric that lies behind the concept of social acceptance. Empirical work
must focus on a subset of both acceptance subjects and acceptance objects. However,
the presentation should help to situate different research approaches and thus clarify
which aspect or constellation of the structure of effects is being focused on.

To underline the relevance of a differentiated interpretation of social acceptance,
we would like to briefly conclude this conceptual discussion with the results of a qualitative study from Japan. From May to September 2016, interviews were conducted
(AT, KT, SN) with a total of 35 members of the general public, of varying ages and
occupations, in Japan [29]. As the purpose of this study was to determine the current
state of willingness of ordinary citizens, the questions did not specify the technical
level of the AVs[1] (for levels of AD see the introduction). Many of the female (20/35)
respondents (17/20) said that they would like to leave driving up to the self-driving
vehicle because they did not feel confident about their driving. In addition, senior
citizens, persons who did not have many opportunities to drive, and other individuals
who seemed to lack confidence in their driving tended to be favourably disposed
toward AVs. However, many respondents expressed unease or concerns, saying they
thought AVs were "*not trustworthy*" and mentioning the problem of determining

---

[1] In order to describe the capabilities of automated driver assistance systems and autonomous
vehicles, regulatory authorities and standardization organizations have introduced systems based
on so-called 'levels'. Most commonly used is the terminology proposed by SAE International in
its standard SAE J3016. Although its 6-level systematics suggests that the higher the level, then
the better, or more advanced the automated system is, this is not exactly the case. The levels only
clarify the division of tasks between the human and the automated system during the operation of
a vehicle. Within the context of AVs, especially the levels 3–5 are of importance. Quite simplified,
level 3 vehicles still rely on human drivers taking over the driving task when the automation requests
while level 4 vehicles are able to perform the entire driving task without human support within a
predefined system of infrastructural and environmental conditions.

responsibility in the event of an accident. Some respondents had separate assessments for their own attitudes and needs, and the needs of society, such as, *"I like driving so I don't need AVs now, but they would be useful for senior citizens and people who cannot drive."* In order to organize these qualitative differences, multiple similar responses were plotted on a graph with support—oppose and like driving—dislike driving axes (Fig. 2) [30]. As Fig. 2 shows, most interviewees were in favour of AVs. However, some respondents did not like AVs very much personally, but were supportive of a society in which AVs had been achieved out of consideration for their convenience and utility to society. There were no opinions in the fourth quadrant of *"liking AVs but being opposed to them (for some reason)"*.

In this way, even some interviewees who did not like AVs personally and did not have a need for them, *"can understand that some people such as senior citizens need them and think that they are useful to society."* This result illustrates one aspect of the diversity of the objects of acceptance and the relationships with the subjects of acceptance, and suggests that the social acceptance of AVs can be interpreted not only by intention to purchase (*"want to buy"*), or intention to use (*"want to use"*), but can be extended to *"willingness to support a society in which AVs have been achieved"*.

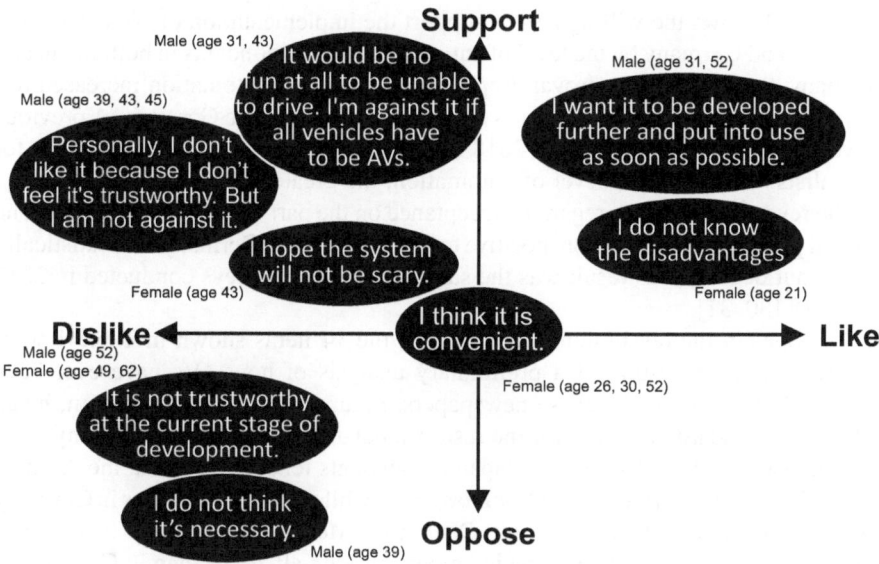

Fig. 2 Results of interviews (n = 35) in 2016 (organized on *like driving—dislike driving* and *support—oppose* axes)

2 Empirical Insights from Quantitative Surveys

In this section we present the results of two comparative surveys. The first was carried out in May 2020 with about 500 people each in Japan and Germany on questions related to the societal acceptance of AVs. Based on the results and experiences of this explorative study, the research team developed a questionnaire for a second, larger-scale survey that was carried out in November 2021.

2.1 Explorative Survey 2020: International Comparison of Willingness to Support AVs

How do members of the general public perceive topics and issues relating to AVs? This section presents examples from an analysis of the results of an online survey conducted among citizens in Japan and Germany. For this exploratory study, 1000 participants (250 each from four regions: Greater Tokyo and Aichi in Japan, and Berlin and North Rhine-Westphalia in Germany) were questioned on the relationship between Tones relating to AVs, and their willingness to support AVs [7, 20].

Figure 3 shows the willingness to support the implementation of AVs in society for Japan and Germany by the level of automation. For respondents in both Japan and Germany, the degree of approval dropped as the level of automation increased (cf. footnote in Sect. 1), but this trend was particularly significant in Germany. A previous survey conducted in Germany in 2018 had similar results. As noted elsewhere, for specialists, the higher the level of automation, the greater the perception of safety, but the reverse is true with regard to acceptance on the part of the general public. The general public in Japan are more positive toward AVs than in Germany to a statistically significant degree. This result was the same in previous surveys conducted in 2017 and 2018 [30, 31].

As Tones in the discussion related to AV, the 14 items shown in Table 1 were selected based on results of a preliminary analysis of how AVs have ever been discussed in articles in Japanese newspapers. Figure 4 shows the histogram, mean values and standard deviation for the results tabulated for Japan and Germany.

In general, as Fig. 4 shows, in Japan respondents tended to choose the "middle way" of choice 3, "*Neither Agree Nor Disagree*", while more respondents in Germany tended to select choice 1, "*Strongly Disagree*". Moreover, as in Fig. 3, in Japan respondents were more likely to be in agreement with all Tones than in Germany.

On closer examination, Tone 5 and Tone 13 are particularly distinctive to examine individual Tone assessments and the determining factors in those assessments. In Germany, 53% of respondents agreed with Tone 5, "*AVs should be introduced to reduce CO_2 emissions by making the entire transport system more efficient*". Conversely, the mean value for Tone 5 as assessed in Japan was not high. As a possible interpretation, we suggest that there may be a high(er) awareness of the environment in Germany, partly as a result of education from a very young age,

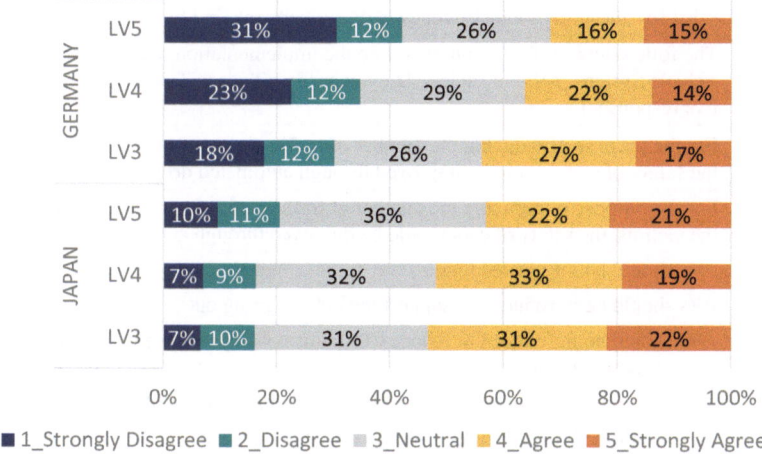

Fig. 3 Degree of support for AVs by the general public in Japan and Germany by level of vehicle automation

and "Stop Climate Change" may function as a "power word" that can get people to respond on an emotional level.

For Tone 13, "*In order to implement AV technology, the government of our country should relax road traffic safety regulations*", more than 40% of German respondents were strongly opposed, selecting choice 1, "*Strongly Disagree*". Conversely, although the mean value for degree of agreement on the part of respondents in Japan was the lowest of all of the 14 Tones, more than 40% of respondents selected choice 3, "*Neither Agree nor Disagree*". Despite the fact that the question text was worded to relax "safety-related" road traffic regulations, many respondents in Japan were hesitant to express their opinions explicitly. This may be because respondents feel that it is a complicated issue whether regulations should be relaxed in order to introduce AVs into society. However, the proportion of respondents in favour of relaxing (safety-related) regulations is much higher than in Germany. This would appear to be the result of a climate and atmosphere in Japan over the past few decades in which any relaxation of restrictions is good. In Japan, reforms such as postal reforms have been promoted based on neoliberal economic policies that advocated "small government", and deregulation is still very often used as a slogan or policy tool in relation to economic policies. As a result, the association that deregulation leads to economic growth may be widely-established and accepted by the public. In Japan, the term "deregulation" might be a "power word" that conjures up a positive impression. In response to other Tones, compared to respondents in Germany, responses in Japan suggest that the perception that "*AVs are needed to stimulate the economy and strengthen international competitiveness*" may have a major impact on support for AVs. Moreover, in every country and city where the survey has been conducted, people with higher confidence in AV technology tended to show greater support for

Table 1 Item list for questions about 14 Tones in discussions relating to AVs

Question:	The following is a discussion regarding the implementation of an Autonomous Vehicle System (AVs) in society. Do you agree with each Tone? (Select from among five responses)
Tone 1	For the purpose of reducing the number of traffic accidents between road vehicles, the safety of cars should be improved through automated driving systems
Tone 2	For the purpose of reducing the number of traffic accidents in which pedestrians are the victims, the safety of cars should be improved through automated driving systems
Tone 3	AVs should be introduced to alleviate traffic congestion
Tone 4	AVs should be introduced to support the elderly going out
Tone 5	AVs should be introduced to reduce CO_2 emissions by making the entire transport system more efficient
Tone 6	AVs should be introduced to support the vulnerable in depopulated areas
Tone 7	AVs should be introduced for effective use of travel time
Tone 8	AVs should be introduced to reduce the cost of transport services such as buses, taxis and trucks
Tone 9	AVs should be introduced to solve the shortage of drivers of transport services such as buses, taxis and trucks
Tone 10	Progress should be made in the social implementation of AV technology to revitalize the domestic economy
Tone 11	Progress should be made in the social implementation of AV technology so that the domestic automobile industry does not lose to international competition
Tone 12	The government of our country should invest to support the social implementation of AV technology
Tone 13	In order to implement AV technology, the government of our country should relax road traffic safety regulations
Tone 14	In order to implement AV technology, the government of our country should conduct AV trials on public roads as soon as possible

AVs. This suggests that confidence in AV "technology" will be extremely important for fostering social acceptance of AVs and that, conversely, if confidence in AV technology is shaken, support for AVs is likely to drop [20].

2.2 Large-Scale Quantitative Survey 2021

In order to better understand expectations and attitudes among the general public towards CAD and related changes in mobility patterns and the regulatory environment, as well as to review and expand the results of the explorative survey presented in Sect. 2.1, we designed a more in-depth quantitative survey for Germany and Japan. The questionnaire used here included the following topic areas: general questions about everyday mobility patterns of the respondents, their own experiences with AVs, individual expectations of the longer-term effects of the use of autonomous road

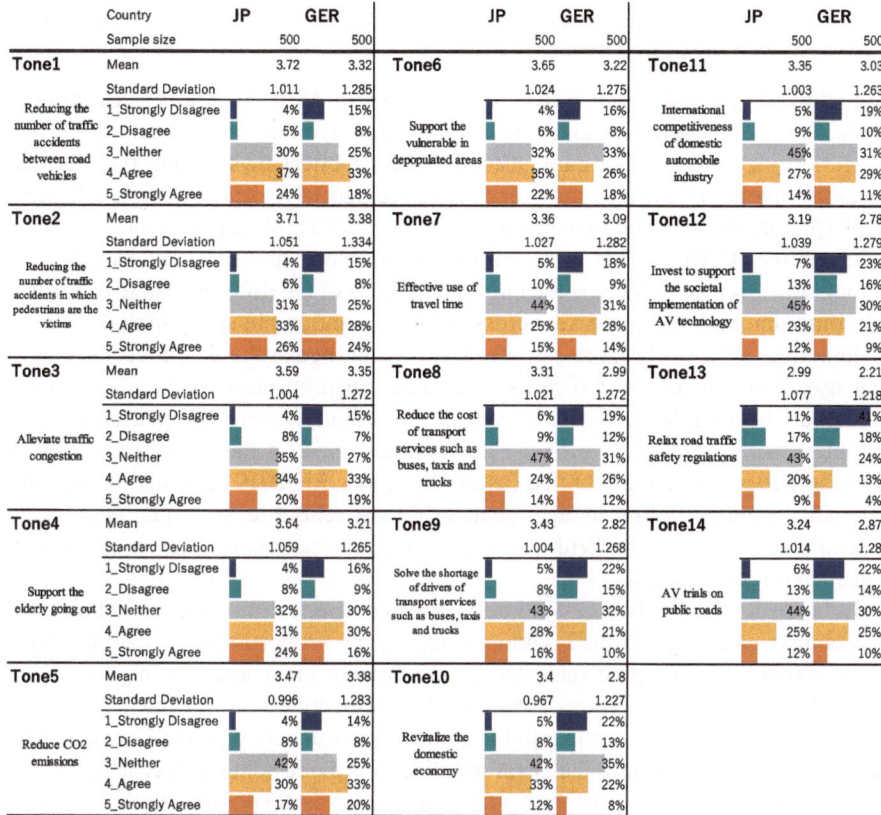

	Country	JP	GER		JP	GER		JP	GER
	Sample size	500	500		500	500		500	500
Tone1	Mean	3.72	3.32	**Tone6**	3.65	3.22	**Tone11**	3.35	3.03
	Standard Deviation	1.011	1.285		1.024	1.275		1.003	1.263
Reducing the number of traffic accidents between road vehicles	1_Strongly Disagree	4%	15%	Support the vulnerable in depopulated areas	4%	16%	International competitiveness of domestic automobile industry	5%	19%
	2_Disagree	5%	8%		6%	8%		9%	10%
	3_Neither	30%	25%		32%	33%		45%	31%
	4_Agree	37%	33%		35%	26%		27%	29%
	5_Strongly Agree	24%	18%		22%	18%		14%	11%
Tone2	Mean	3.71	3.38	**Tone7**	3.36	3.09	**Tone12**	3.19	2.78
	Standard Deviation	1.051	1.334		1.027	1.282		1.039	1.279
Reducing the number of traffic accidents in which pedestrians are the victims	1_Strongly Disagree	4%	15%	Effective use of travel time	5%	18%	Invest to support the societal implementation of AV technology	7%	23%
	2_Disagree	6%	8%		10%	9%		13%	16%
	3_Neither	31%	25%		44%	31%		45%	30%
	4_Agree	33%	28%		25%	28%		23%	21%
	5_Strongly Agree	26%	24%		15%	14%		12%	9%
Tone3	Mean	3.59	3.35	**Tone8**	3.31	2.99	**Tone13**	2.99	2.21
	Standard Deviation	1.004	1.272		1.021	1.272		1.077	1.218
Alleviate traffic congestion	1_Strongly Disagree	4%	15%	Reduce the cost of transport services such as buses, taxis and trucks	6%	19%	Relax road traffic safety regulations	11%	41%
	2_Disagree	8%	7%		9%	12%		17%	18%
	3_Neither	35%	27%		47%	31%		43%	24%
	4_Agree	34%	33%		24%	26%		20%	13%
	5_Strongly Agree	20%	19%		14%	12%		9%	4%
Tone4	Mean	3.64	3.21	**Tone9**	3.43	2.82	**Tone14**	3.24	2.87
	Standard Deviation	1.059	1.265		1.004	1.268		1.014	1.28
Support the elderly going out	1_Strongly Disagree	4%	16%	Solve the shortage of drivers of transport services such as buses, taxis and trucks	5%	22%	AV trials on public roads	6%	22%
	2_Disagree	8%	9%		8%	15%		13%	14%
	3_Neither	32%	30%		43%	32%		44%	30%
	4_Agree	31%	30%		28%	21%		25%	25%
	5_Strongly Agree	24%	16%		16%	10%		12%	10%
Tone5	Mean	3.47	3.38	**Tone10**	3.4	2.8			
	Standard Deviation	0.996	1.283		0.967	1.227			
Reduce CO2 emissions	1_Strongly Disagree	4%	14%	Revitalize the domestic economy	5%	22%			
	2_Disagree	8%	8%		8%	13%			
	3_Neither	42%	25%		42%	35%			
	4_Agree	30%	33%		33%	22%			
	5_Strongly Agree	17%	20%		12%	8%			

Fig. 4 Attitudes towards societal and/or policy goals linked to AVs in public debates in Japan and Germany

vehicles, perceptions of different AD use cases, expectations regarding future framework conditions and regulations in connection with the increased use/deployment of autonomous road vehicles, and the perceived potential of autonomous road vehicles to fulfill individual mobility needs. The items were collectively developed by a Japanese-German research team and presented to the respondents in their respective languages.

Because the general population was to be surveyed within the framework of this study, we expected that a substantial number of those questioned were not familiar with the concept of AD. Therefore, the questionnaire opened with a short text explaining it:

Worldwide, work is underway to develop autonomous vehicles for road traffic. These vehicles are controlled by a computer. They should be able to travel at least as safely and flexibly as today's vehicles. Because they do not need a human driver, they no longer have a steering wheel or pedals. We are particularly interested in how such vehicles could change our everyday lives in the future. Your answers are therefore very valuable to us—even if you feel you don't know much about them so far

Fieldwork in both countries was conducted in November 2021. Because of methodological challenges and limited resources, we had to apply different data collection methods. In Germany, a total of 2,001 interviews among the German-speaking population aged 16 and over were conducted, of which 1,001 were telephone (CATI) interviews in a dual-frame approach, with 50% mobile phone interviews and 1,000 online (CAWI). This mixed-mode methodology and the large sample size, among other things, reduces some empirical effects of the different usage patterns of telecommunication technologies across different demographic subpopulations while maintaining sufficiently large sample sizes in contingency tables, especially those that use dominant mode choice as a variable. Disproportionalities arising in the course of the sample design or survey implementation were compensated for by complex, iterative weighting of the net sample based on the latest data from the German Federal Statistical Office (as of December 31, 2019) and taking the characteristics of household size, age, gender, highest school-leaving qualification and federal state into account. With respect to these characteristics, the sample can be considered representative.

In Japan, a total of 1,058 CAWI interviews were used for this analysis. The sample was composed using stratified random sampling by proportional allocation relative to three socio-demographic variables: gender (two groups), age (seven groups), and place of residence (eight groups), based on their respective demographic composition ratio. This limits the full comparability of the Japanese data with the German study, but in our view is still sufficiently informative to provide some deeper insights.

A complete presentation of the results of the survey is not possible within the space available here. We therefore limit ourselves to an initial analysis of the topic areas, 'Individual expectations on the longer-term effects of the use of autonomous road vehicles' (Question Set 4), 'Perceptions of different AD use cases' (Question Set 5), and 'Attitudes towards future framework conditions and regulations in connection with the increased use / deployment of autonomous road vehicles' (Question Set 6). This is mainly because we think these will provide helpful insights into public expectations regarding the design of future AVs and the mobility services that use them, and hence will be of particular interest in the context of social acceptance.

2.2.1 Longer-Term Effects of AV Deployment

In order to evaluate individual expectations of the longer-term effects of the widespread use of autonomous road vehicles and to capture how respondents assess the importance of the "four promises" of CAD, survey participants were asked: *"Please imagine that in the future there would be autonomous road vehicles that would be able to participate in public road traffic just as independently as vehicles with human drivers do today. What would you expect from such a development in the longer term?"*, on an 11 point Likert scale with 0 (= *"I would not expect that in any case*") and 10 (= "*I would definitely expect that*") as verbal labels for the endpoints. Thirteen items were offered (Table 2). The core statistical measures for all items are shown in Table 3.

The range of both the statistical average (between 4.78 and 6.58 for Germany and 5.03 and 6.29 for Japan) and the median values (between 5 and 7) suggest that, in general, the expectations of longer-term impacts in both countries are rather muted. Among the presented items, none was seen as outstandingly likely or extremely polarizing. Values for standard deviation and the Box2-values indicate that the answers in the German sample tended to be partially more definitive, while the answers in the Japanese sample have a stronger tendency towards the middle of the scale.

Table 2 Item list for Question Set 4: individual expectations of the longer-term effects of the use of autonomous road vehicles

No.	Short	Full item wording
4.1	Reduced number of accidents	The number of traffic accidents will decrease
4.2	Reduced severity of accidents	The severity of traffic accidents (the number of people killed and seriously injured in them) will decrease
4.3	Smoother road traffic	Road traffic will run more smoothly overall, and there will be fewer traffic jams
4.4	Fewer parked cars	There will be fewer parked cars than today
4.5	Children will travel independently	Children will travel more distances independently, i.e., without being accompanied by their parents or other adults
4.6	Elderly people will travel independently	Elderly people and people with limited mobility will make more trips independently
4.7	Improved public transport services	Public transport services will improve, especially in less densely-populated areas (such as on the outskirts of cities, in small towns and in rural areas)
4.8	Mobility services cheaper overall	Mobility services will become cheaper for customers overall
4.9	More traffic on the roads	There will be more traffic on the roads
4.10	Economic competitiveness will be strengthened	The competitiveness of the German (for survey in Germany)/ Japanese (for survey in Japan) economy will be strengthened as a result
4.11	Climate gas emissions will be reduced	Climate gas emissions from transport will be reduced
4.12	Time in traffic used productively	I will use the time I spend on the road for productive purposes (such as working, doing homework, or attending meetings)
4.13	Mobility services cheaper for me	Mobility services will become cheaper for me overall

Table 3 Core statistical measures for Question Set 4: expectations regarding longer-term effects of AV deployment among the German (left column) and Japanese (right column) populations

Item	Germany					Japan				
	Averages			Top2-Box		Averages			Top2-Box	
	ArMean	StdDev	Med	Top (%)	Bottom (%)	ArMean	StdDev	Med	Top (%)	Bottom (%)
4.1	5,95	3,14	6	23	12	5,71	2,60	6	14	7
4.2	5,99	3,09	6	23	11	5,73	2,64	6	14	8
4.3	6,33	3,10	7	28	11	5,52	2,49	6	11	7
4.4	6,11	3,19	7	25	12	5,49	2,54	6	11	7
4.5	4,86	3,02	5	12	16	5,03	2,55	5	8	10
4.6	6,58	3,03	7	30	9	6,29	2,47	7	20	4
4.7	6,16	3,06	7	24	10	6,17	2,40	6	16	5
4.8	5,10	3,12	5	15	16	5,69	2,49	6	12	7
4.9	4,78	2,93	5	11	13	5,24	2,32	5	8	6
4.10	5,50	3,00	5	16	12	5,52	2,37	6	9	6
4.11	6,10	3,21	7	26	12	5,64	2,60	6	14	8
4.12	5,54	3,35	6	22	17	5,22	2,63	5	10	10
4.13	4,88	3,18	5	14	19	5,48	2,49	5	10	8

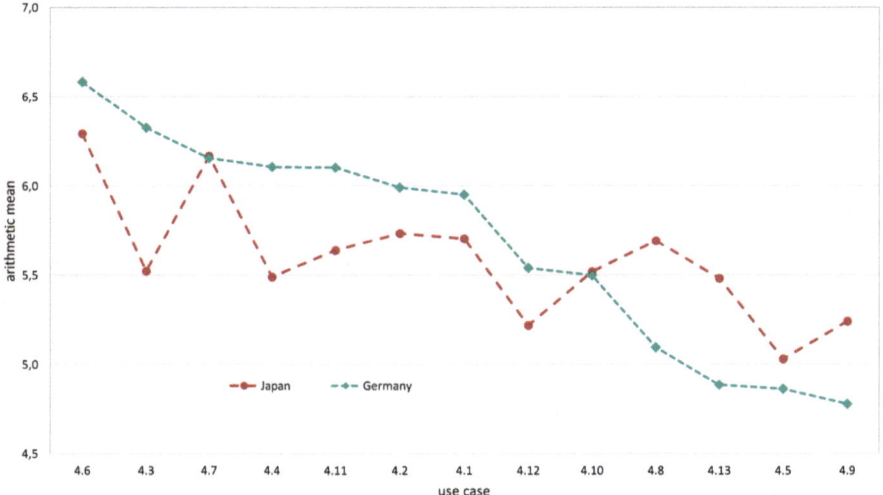

Fig. 5 Mean values for Question Set 4 for Germany and Japan, presented in descending order of German items

This tendency can also be seen in Fig. 5, where the x-axis represents the approval ratings of items within the German sample in descending order. In addition, respondents in Japan are more optimistic about cost reductions for mobility services stimulated by the introduction of AVs, while respondents in Germany are more optimistic about the implications for stationary and flowing traffic.

As shown in Fig. 6, about one third[2] of the respondents in Germany and one quarter in Japan were quite sure that the introduction of AVs will bring improvements with regard to traffic safety, measured both as number (Q4.1) and severity (Q4.2) of traffic accidents. About 15% in Germany and 12% in Japan remained rather skeptical.

Asked about potential mobility improvements for groups who are not able to drive themselves, almost half of the German and one third of the Japanese respondents expected that AVs will enable elderly people and people with limited mobility to make more trips independently (Q4.6). They were rather unsure and divided with respect to the independent use of AV by unaccompanied children (Q4.5).

The picture remains rather undecided in both countries, and even shows an element of polarization in Germany, with regard to two service implications of AV deployment: the opportunity to use time in traffic for productive purposes (Q4.12), and the reduction of mobility cost (Q4.8).

With respect to improvements of transportation efficiency, about 45% of German respondents expected that the introduction of AVs will contribute to a reduction of the number of traffic jams and a smoother traffic flow (Q4.3). This is in line with the result that a little less than 40% of respondents in Germany expected that the introduction of AVs will lead to a reduction of climate gas emissions from transport

---

[2] When referring to distribution data, we report on the Top-3- (8–10) or Bottom-3- (0–2) Boxes.

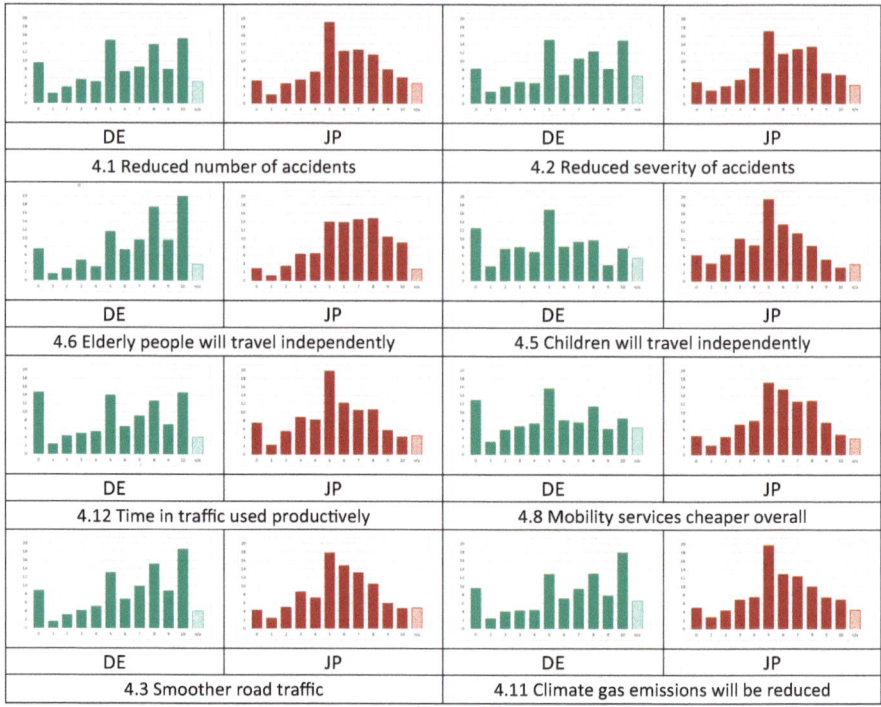

Fig. 6 Selection of histograms of the responses to items in Question Set 4: expectations regarding longer-term effects of AV deployment

(Q4.11). In Japan, respondents were a little more hesitant here: ca. 20% shared these expectations.

The dataset shows some variations with respect to sociodemographic factors. As a tendency (but not a rule), the expectations are slightly higher among younger respondents, people with a higher level of completed education, and men. The averages for the male subpopulation differ by 0.5 or more (when compared to the female subpopulation) for Q4.1 to 4.5 (men higher) within the German dataset, while the differences are smaller than 0.3 for all other items. The differences are somewhat smaller within the Japanese dataset, but follow similar trends with few exceptions (Fig. 7). So, e.g., the differences between men's and women's expectations regarding economic effects (Q4.10) and productive time use (Q4.12) were much larger in Japan than in Germany. There are only one (Germany) or two items where expectations are higher among women than among men: climate gas emissions from transport will be reduced (Q4.11—both countries), and public transport services will improve, especially in less densely-populated areas (Q4.7—Japan only).

Correlation analysis for the German dataset showed that the Pearson correlation coefficient (PCC) for any pair of 12 of the 13 items was 0.4 or higher (significant

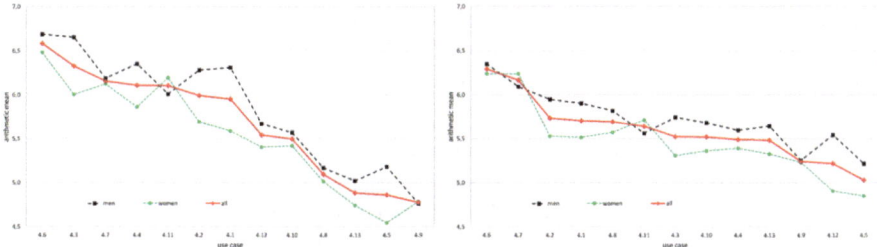

Fig. 7 Mean values for Question Set 4, presented in descending order and differentiated by gender, for Germany (left) and Japan (right)

at the 0.01 level (2-tailed)), generally indicating a moderate to strong positive relationship between two variables. The only exception is Q4.9 (*"There will be more traffic on the roads"*), which does not appear to correlate with any of the other 12 items. This observation is also confirmed by an Exploratory Factor Analysis (EFA) where the same 12 items load on one factor. This suggests that most of the current expectations regarding longer-term effects of AD might be shaped by a single latent factor. Correlation analysis with the Japanese data set provides a basically similar picture. Here, the Pearson coefficient showed significant (at the 0.01 level (2-tailed)) correlation for all 13 items. Its values were, like in the German case, 0.4 or higher. Q4.9 again proved to be the exception. Other than in the German case where there was no correlation, it is correlated to the other 12 items but the PCCs were in the range of 0.3 and hence somewhat smaller than among the other items.

2.2.2 Framework Conditions and Regulatory Adaptation

It is well-known both from the innovation research literature as well as from policy practice that the large-scale deployment of new, presumably transformative technologies requires adaptations of the regulatory framework. In highly-regulated fields (like road transportation) this is usually the case before the introduction of a new technology. Non-acceptance of a technology might be rooted in skepticism about changes in the framework conditions and the perceived impact on one's own lifestyle, rather than in the characteristics of the technology itself.

In an attempt to capture some of these effects and to gain further insights into the expectations regarding future framework conditions and regulations in connection with the increased use/deployment of AV, participants were asked: *"To make such a development toward autonomous driving possible, some framework conditions of today's traffic might have to be changed. Assuming that would include the following changes: Would you be more likely to welcome or more likely to oppose them?"*. Analogous to Question Set 4, an 11 point Likert scale was offered, in this case with 0 (= *"I would definitely reject this"*) and 10 (= *"I would definitely welcome this"*) as verbal labels for the endpoints. The following thirteen items were presented (Table 4); the core statistical measures for all items are shown in Table 5.

Table 4 Item list for Question Set 6: attitudes towards future framework conditions and regulations in connection with the increased use/deployment of autonomous road vehicles

No.	Short	Full item wording
6.1	Financial support to purchase AVs	The government should provide financial support for private individuals to purchase autonomous vehicles
6.2	Type approval framework should be relaxed	The existing regulatory framework for the type approval of motor vehicles should be relaxed to make it easier to offer new mobility services with autonomous vehicles
6.3	Liability with manufacturers	If autonomous vehicles are involved in an accident, their manufacturers should assume liability for damages
6.4	Liability with owners	If autonomous vehicles are involved in an accident, their owners should assume liability for damages
6.5	DPR should be relaxed	Data protection regulations (DPR) should be relaxed
6.6	Users can intervene in AD	Users should also be able to intervene in autonomous driving if accidents are imminent
6.7	AVs only in segregated lanes	Autonomous vehicles should only be allowed to drive in their own lanes, which must be structurally separated from other road traffic
6.8	AVs immediately identifiable	It should be easy for every road user to recognize at all times whether a vehicle is driving autonomously
6.9	AVs may violate traffic rules	Autonomous vehicles should be allowed to violate traffic rules if this could prevent accidents
6.10	AVs tested on public roads	Autonomous vehicles should be allowed to be tested in transparent field trials on public roads
6.11	Citizens involved in field trials	Citizens should be involved in planning and conducting field trials of autonomous vehicles
6.12	Generous testing opportunities for private sector	Private mobility providers should be given generous opportunities to test new services with autonomous vehicles
6.13	AVs should adapt to VRU	Autonomous vehicles should drive carefully when they perceive children or elderly people (vulnerable road users, VRU) in their vicinity

Compared to Question Set 4, the responses were much more diverse and, in some cases, very definitive. This is indicated by the fact that statistical averages range from 3.96 to 8.44 for Germany and between 4.87 and 7.43 for Japan. Median values lie between 4 and 10 for Germany and 5 and 8 for Japan. Q6.3 and Q6.4 were removed from further analyses since we found that the way the questions were presented in the survey does not meaningfully capture the actual regulatory situation. As was the case for Q4, values for standard deviation and the Box2-values signal that the answers in the German sample tended to be partially more definitive, while the answers in the Japanese sample have a stronger tendency towards the middle of the scale.

This tendency can also be seen in Fig. 8, where the x-axis represents the approval ratings of items within the German sample in descending order. As a tendency, regulatory measures that would improve (perceived or actual) safety (6.13, 6.6., 6.8)

Table 5 Core statistical measures for Question Set 6: attitudes toward changes in framework conditions and regulatory adaptations among the German (left column) and Japanese (right column) populations

Item	Germany					Japan				
	Averages			Top2-Box		Averages			Top2-Box	
	ArMean	StdDev	Med	Top (%)	Bottom (%)	ArMean	StdDev	Med	Top (%)	Bottom (%)
6.1	5,70	3,50	6	26	17	5,95	2,44	6	14	5
6.2	5,20	3,15	5	15	16	5,72	2,42	6	12	6
6.3	7,33	2,85	8	41	5	6,52	2,34	6	21	3
6.4	5,55	3,62	5	27	19	6,06	2,52	6	16	6
6.5	3,96	3,39	4	11	32	4,87	2,60	5	8	11
6.6	7,88	2,80	9	54	5	7,03	2,39	7	30	3
6.7	5,65	3,41	6	25	16	6,24	2,43	6	19	4
6.8	8,01	2,73	9	57	5	6,85	2,31	7	27	3
6.9	5,96	3,35	7	25	15	6,03	2,39	6	14	5
6.10	6,74	3,04	7	34	9	6,85	2,31	7	25	3
6.11	7,29	2,77	8	39	6	6,51	2,24	7	19	3
6.12	6,27	2,91	7	22	10	6,30	2,24	6	16	3
6.13	8,44	2,40	10	63	3	7,43	2,33	8	38	2

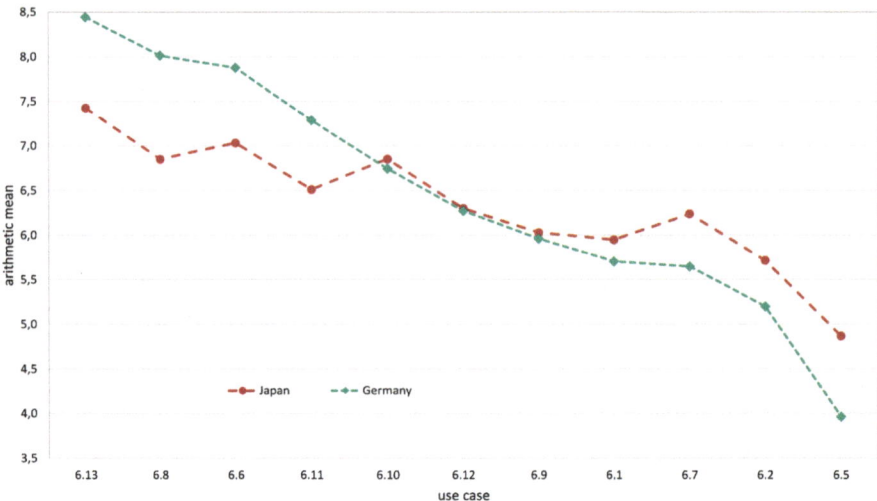

Fig. 8 Mean values for Question Set 6 for Germany and Japan, presented in descending order of German items

are most welcome in both countries, and even stronger in Germany than in Japan, while measures that aim at the relaxation of regulatory frameworks (like 6.5 or 6.2) are those that were most opposed.

The most unambiguous results are related to the situational behavior of AVs in road traffic. A majority of the survey participants (75% in Germany, 53% in Japan) would support that AVs should drive carefully when they perceive children or elderly people in their vicinity (Q6.13) and that they should be easily identifiable as driving autonomously at all times and for every road user (Q6.8). While meeting the first expectation would create substantial challenges for technology developers and traffic management, the second could be met rather easily by making respective indicators a part of the design criteria and the type approval process of AV.

About two thirds of the German respondents and 45% of the Japanese participants would welcome the ability to intervene in AD if accidents were imminent from their perspective (Q6.6). Allowing AVs to violate traffic rules if this could prevent accidents (Q6.9) was seen to be much more controversial, but with a slightly supportive tendency.

As shown in Fig. 9, roughly half of all German respondents agreed that testing strategies for AVs should support transparent field trials on public roads (Q6.10), and even more would welcome citizens' involvement in planning and conducting these field trials (Q6.11). In Japan, the picture was reversed. The overall agreement with field testing was even higher than in Germany, but the support for citizens' participation was remarkably lower. We hypothesize that this might be rooted in different political traditions in the two countries (for further discussion, see chapter "Governance, Policy and Regulation in the Field of Automated Driving: A Focus on Japan and Germany").

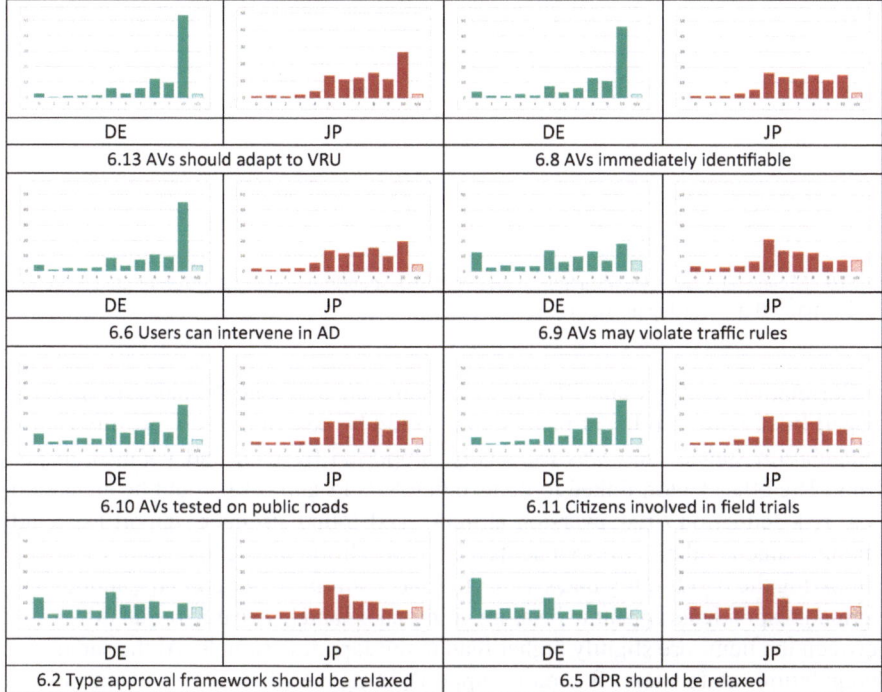

Fig. 9 Selection of histograms of the responses to items in Question Set 6: attitudes toward changes in framework conditions and regulatory adaptations

Finally yet importantly, two views on the overall regulatory framework should be mentioned: Respondents were rather undecided whether type approval regulations should be relaxed (Q6.2) in order to allow for a faster diffusion of AVs. This might be due to that fact that most participants possibly do not know that these regulations exist, and which role they play for traffic and product safety. Asked about data protection rules (Q6.5), respondents were much clearer. For both countries, this was the only item within Question Set 6 where the mean fell below the center of the answer scale. For German respondents the reaction was much stronger than in Japan, and even somewhat polarized: A little less than 40% were rather opposed to relaxations of these rules, while 20% supported easing them. In Japan, responses showed a stronger tendency towards the scale center, with 15% being in favor of relaxing data protection rules and 18% opposing it.

When comparing the answer sets between male and female respondents in both countries (Fig. 10), the differences overall are rather small, but women appear to be more supportive of measures improving (perceived or actual) safety, and more opposed to the relaxation of regulations than men. In both countries, the largest difference between male and female respondents occurs for Q6.7 (*"Autonomous vehicles should only be allowed to drive in their own lanes, which must be structurally separated from other road traffic"*).

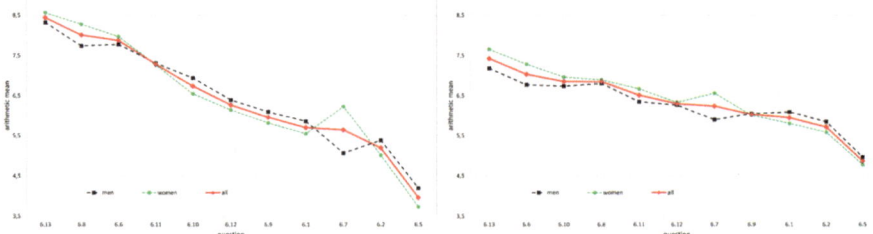

Fig. 10 Mean values for Question Set 6, presented in descending order and differentiated by gender, for Germany (left) and Japan (right)

Correlation analysis and EFA for Germany suggest that the answers might be shaped by at least two latent factors. One could be described as a 'pro-innovation/ deregulation stance' and was especially supported by men and younger respondents. The other factor, linked to items 6.6, 6.8, 6.11 and 6.13, could be understood as a 'risk reduction / risk aversion attitude' and found stronger support especially among women and older respondents. The latter differences can be identified when comparing the two country datasets. Both follow similar general trends, although the German respondents appear to be more opinionated and hence the overall differences between the items are slightly higher than in the Japanese sample. At the same time, deregulation-related items find more support among men than among women in both countries, while safety-related items find more support among women than among men. In both countries, item Q6.13 ("*Autonomous vehicles should drive carefully when they perceive children or elderly people in their vicinity*") gained strongest support among all items, while the relaxation of data protection rules (Q6.5) found least support overall. The graphs also confirm the relatively higher support for public participation in field trials with AV (Q6.11) in Germany. Beyond this, support for public testing was higher among women than among men in Japan, while in Germany support was higher among men.

2.2.3   Use Cases

A third Question Set was dedicated to the presumed wellbeing of respondents in the course of using different mobility options enabled by CAD technologies: "*Imagine that in the future there would be autonomous road vehicles that would be able to participate in public traffic just as independently as vehicles with human drivers do today. In which constellation would you feel comfortable driving such a vehicle?*". In analogy to the other two Question Sets discussed here, an 11 point Likert scale was offered, in this case with 0 (= "*I would not feel at all comfortable*") and 10 (= "*I would definitely feel comfortable*") as the verbal labels for the endpoints. Seven different use cases were offered, varying by occupancy, type of mobility service and traffic environment (Table 6). These seven cases cover the new mobility services commonly discussed today for urban transport (robotaxi, automated mini-shuttle for ridesharing, own vehicle) and add further variations and options. The core statistical measures for all items are shown in Table 7.

The first thing to notice is that both the mean and the median are close to the center of the distribution and the differences are rather small overall, i.e., there was no pronounced preference for or aversion to individual use cases in the population as a whole. This is also made clear by the graphical representation of the distributions, of which only a selection can be presented here for reasons of space.

At the same time, as shown in Fig. 11, it can be seen that in certain cases even small changes in the imagined system design lead to measurable changes in wellbeing, such as the introduction of a teleoperator in a robotaxi (5.3 and 5.4), or the change from public bus to streetcar (5.6 and 5.7). The follow-up hypothesis is therefore that wellbeing in new mobility services (and thus their acceptance) is not determined solely by the automation concept and the degree of automation (level), but that other design and service factors must also be taken into account.

Further insights are provided by differentiating the response behavior according to sociodemographic characteristics, such as gender and age. Regardless of the use case, and in both countries, women state that they would feel comfortable in automated means of transportation significantly less frequently than men (Fig. 12).

A substantially more complex outcome emerges from a comparison between age groups. For German respondents, the picture is rather clear-cut: independent of the use case, subjective wellbeing while using automated transport is significantly lower among the older population than among younger people (Fig. 13). In Japan, the differences among age groups are remarkably smaller than in Germany. Regardless of the use case, among the youngest respondents (16–29) subjective wellbeing is

Table 6 Item list for Question Set 5: attitudes towards future framework conditions and regulations in connection with the increased use/deployment of autonomous road vehicles

No.	Short	Full item wording
5.1	Private AV on highway	Alone in my private autonomous vehicle on a highway driving at the speed of the advisory speed limit
5.2	Private AV in city traffic	Alone in my private autonomous vehicle in city traffic
5.3	Robotaxi in city traffic	Alone in a hired autonomous vehicle in city traffic
5.4	Monitored robotaxi in city traffic	Alone in a hired autonomous vehicle in city traffic, where the journey is constantly monitored by a tele-operator
5.5	Mini-Shuttle ridesharing	Together with two to five other passengers in an autonomous mini-bus in city traffic
5.6	Autonomous city bus	In a half-full autonomous bus the size of today's city buses in urban traffic
5.7	Autonomous streetcar	In a half-full autonomous streetcar in urban traffic

Table 7 Core statistical measures for Question Set 5: Subjective wellbeing when using different AV-based mobility services among the German (left column) and Japanese (right column) population

Item	Germany						Japan					
	Averages			Top2-Box			Averages			Top2-Box		
	ArMean	StdDev	Med	Top (%)	Bottom (%)		ArMean	StdDev	Med	Top (%)	Bottom (%)	
5.1	4,76	3,51	5	17	26		4,96	2,71	5	8	14	
5.2	5,26	3,47	5	22	21		5,13	2,67	5	9	12	
5.3	4,86	3,36	5	16	23		4,89	2,63	5	7	13	
5.4	5,07	3,27	5	16	20		5,03	2,63	5	9	13	
5.5	5,10	3,14	5	14	18		5,22	2,54	5	9	10	
5.6	5,34	3,14	6	16	16		5,20	2,52	5	8	10	
5.7	5,66	3,19	6	20	15		5,47	2,58	6	11	9	

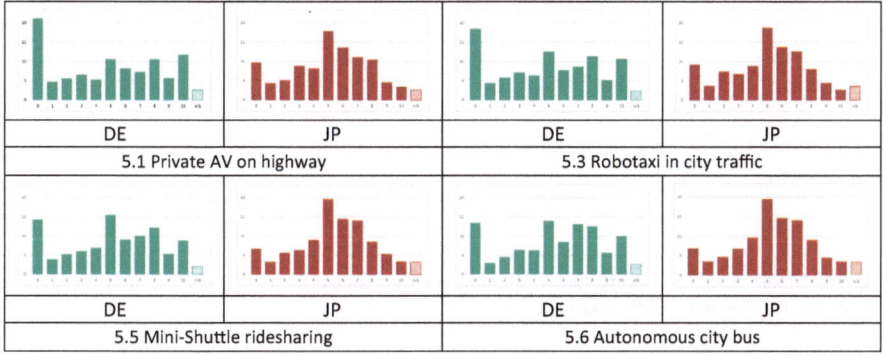

Fig. 11 Selection of histograms of the responses to items in Question Set 5: subjective wellbeing for hypothetical cases of different future AV-based mobility services

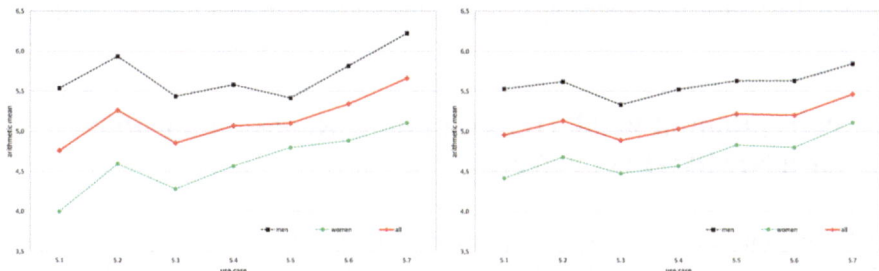

Fig. 12 Mean values for subjective wellbeing for different use cases, differentiated by gender, for Germany (left) and Japan (right)

higher than in the average population, while it is lower among the age group 45–59. For older respondents (60 and above), subjective wellbeing while using collective automated mobility service options (5.5–5.7) is higher than in the average population, while it is lower for individual mobility service options.

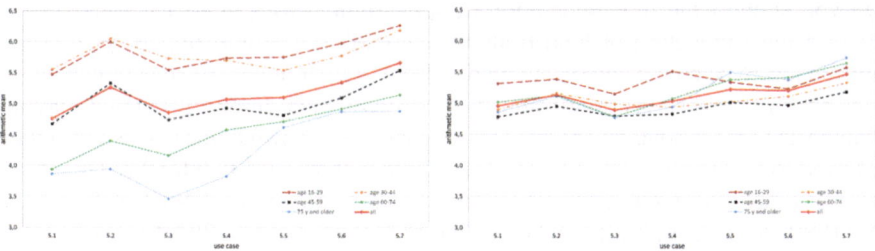

Fig. 13 Mean values for subjective wellbeing for different use cases, differentiated by age group, for respondents in Germany (left) and Japan (right)

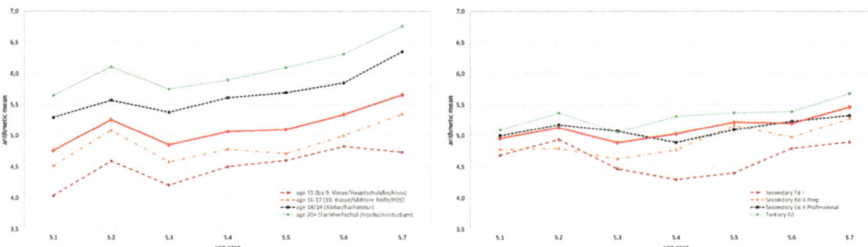

Fig. 14 Mean values for subjective wellbeing for different use cases, differentiated by level of education, for Germany (left) and Japan (right)

Subjective wellbeing for different use cases of AVs appears to increase with increasing levels of formal education, in both countries (Fig. 14). The most important difference is that German respondents with completed academic (tertiary) education judged their subjective wellbeing considerably higher than their peers in Japan.

3   Perceptions of Field Trials on Public Roads—Indicators for a NIMBY-Like Effect?

Field trials on public roads are indispensable to bring forward the integration of AVs into mobility systems. Such tests have been conducted for several years in both Japan and Germany in many different contexts. For many citizens, AVs are becoming more concrete and tangible only through such field trials. In this section, we report on how such tests are perceived by citizens, and how they can influence the perception of AVs.

3.1   The "NIMBY" Issue in Field Trials

Do you support the implementation of self-driving buses in society, and do you support the conducting of field trials to that end? If you are supportive, will you allow a test to be conducted on the road in front of your home? Even people who say they are in favour of AVs and want these systems to be further promoted may not want tests to be conducted on the road in front of their own homes. This problem is referred to by the phrase Not In My Back Yard, abbreviated to NIMBY. The phrase is said to have been coined by Walter Rogers of the American Nuclear Society in the 1980s to describe people who wanted to receive the benefits of nuclear power generation while at the same time being opposed to the building of nuclear power plants in their area [5]. Does the NIMBY problem exist with regard to AV field trials? Based on this research question, in May 2020 an online survey was conducted with

1,000 members of the general public,500 in Japan and 500 in Germany. The results were as follows [28].

Figure 15 shows a graph with measurements of the NIMBY level in Japan and Germany with regard to, (a) willingness to support the implementation of AVs in society, and (b) willingness to support a field trial being conducted in front of the respondent's own home. If, for example, the respondent expressed support for the implementation of AVs in society and assigned a score of 4 or 5, and yet was opposed to an AV test being conducted in front of their own home, assigning a score of 2 or 1, this would be considered to be an example of a NIMBY attitude and would be coloured orange. Conversely, if the respondent was opposed to the implementation of AVs in society but would support a field trial being conducted in front of their own home, that would be considered to be an example of a YIMBY (Yes In My Back Yard) attitude and would be coloured green. There is room to debate how to interpret the term YIMBY. It may mean permitting the introduction of AVs into a limited area such as in front of one's home, but being against introduction throughout the whole of society. Or respondents may have the following attitudes: They do not support the implementation of AV for now, because they do not yet like, trust, or know the current plan and technology. But by doing the field trial in front of their house, they can learn more about AV, and maybe even have a chance to get involved in transport planning in the way that they want. It could also be interpreted as the following attitudes: They do not support the implementation of AV for now, because they do not like, trust, or know the current plan and technology yet. But by doing the field trial in front of their house, they can know more about AV, and maybe even have a chance to get involved in transport planning in the way that they want.

Moreover, the views shown in the section running diagonally from upper left to lower right were considered by the researchers to be consistent either in favour of or opposed to both implementation in society and testing in front of one's home. The

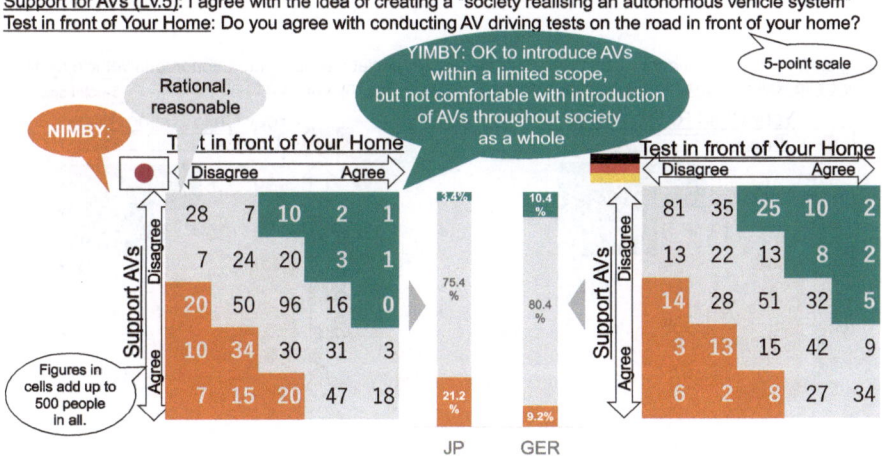

Fig. 15 Comparison of NIMBY factor for AV field trials in Japan and Germany

vertical bar graphs in the centre compare the percentages of NIMBY, YIMBY and 'consistent' responses in Japan and Germany. These graphs show that the proportion of NIMBY attitudes in Japan is at least double the proportion in Germany. While many German respondents gave consistent responses, in Japan, with its high proportion of NIMBY responses, care is needed, as it is possible that the number of people in opposition will increase when concrete plans for field trials on public roads are established.

Next, the same analysis was conducted, but the measure was changed so that instead of willingness to support the field trial held in front of one's home, one's own child would be placed unaccompanied onboard a Level 5 AV (when it becomes available). Figure 16 shows the results. Compared to Fig. 15, the proportion of NIMBY responses increased approximately 1.6 times for Japanese respondents and 2.5 times for German respondents. In addition, the proportion of YIMBY responses from both Japan and Germany dropped to the 1% level. These results may show that only when respondents were asked about their own children did they begin to think about AVs as something that affected them personally. In other words, it is necessary to recall that the responses to the questionnaire survey about willingness to support AVs, intention to use and intention to purchase are only stated preferences (SP) in a hypothetical situation. These may be different from the action taken in the event that AVs are actually implemented, and while we currently have to depend on SP to measure social acceptance, there are limits to its utility.

In the study by Tanaka et al. [28] to determine the characteristics of people who tended to provide NIMBY responses, the difference between a Level 5 response in Fig. 16 for willingness to support AVs and the degree of acceptance of the respondent's own child being placed unaccompanied on board the test vehicle as a passenger was used as the NIMBY factor (dependent variable), and the impact on the NIMBY value of sex, age, presence of children 12 or younger, vehicle ownership, travel behaviour, trust, perception of risks with regard to AD (fear, unknowability) and other factors were analysed (multiple linear regression analysis). The authors

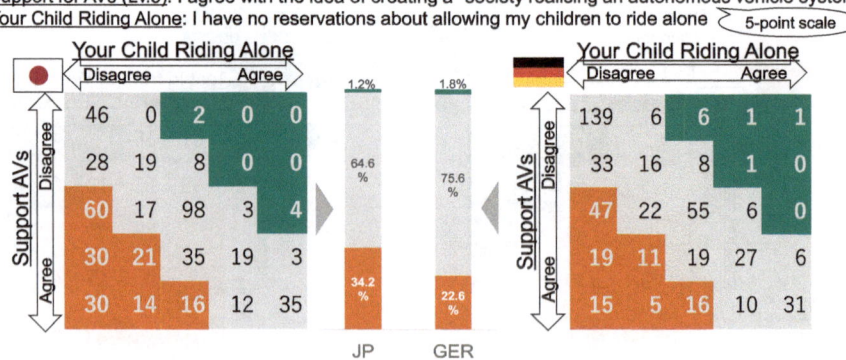

Fig. 16 Comparison of NIMBY factor for a scenario assuming one's own child riding unaccompanied as a passenger on AVs in Japan and Germany

(Tanaka et al. [28]) have previously reported that respondents with a high NIMBY factor have insufficient knowledge of AVs (high degree of unknowability with regard to risk perception) and think that AVs are frightening (high degree of fear with regard to risk perception). Acquainting people with the mechanism and limitations of AVs will be important to decrease the tendency for people to take a NIMBY approach and possibly impede the implementation of field trials.

3.2 Is Social Acceptance of AVs Different After Field Trials? A Study from Japan

In the following we report a study that was carried out only in Japan, where in 2017, field trials of AVs on public roads began. AT assisted in the opinion survey of social acceptance conducted with onboard monitors, and local residents who participated in "Field trials of Automated Driving Service in Hilly and Mountainous Areas Based at Michi-no-Eki etc." conducted by the Ministry of Land, Infrastructure, Transport and Tourism [13, 14]. Of the region-designated type and open-bid type field trial regions, Fig. 16 shows the tabulated results for 1,346 persons in nine regions for whom data could be obtained before and after the test. The four indicators (5 point Likert scale, $1 = $ negative, $5 = $ positive) that were measured were:

(1) Willingness to support AVs (whether the person was in favour of introducing public transport using self-driving vehicles to the region).
(2) Intention to use AVs (whether the person wanted to use public transport using AVs in the future).
(3) Trust in AV technology (does the person think the automated driving technology is trustworthy?).
(4) Trust in AV administration (does the person trust the administration or company building the social mechanisms relating to automated driving?).

Figure 17 shows there was a positive change in all four indicators after the test compared to before the test for the onboard monitors participating in the field trials. In contrast, for the local residents who did not participate in the field trial, only the indicator for trust in AV technology changed to positive. Compared to those who did not participate in the field trials, even before their participation the monitors were clearly more positive about the introduction of AVs into the local public transportation network, and already had a higher intention to use them in the future (cf. Sect. 1). The field trials can be said to have influenced test participants to be more supportive of a society in which public transport using AVs has been achieved.

In addition, Fig. 18 shows the results of path analysis of the relationship of these four psychological indicators [14]. As these results show, not only trust in AV technology but trust in the administration or company building the social mechanisms for AVs has an impact on the intention to use AVs; and the intention to use AVs has an impact on the willingness to support AVs. The development of AV technology is of course important, but these results also indicate the importance of

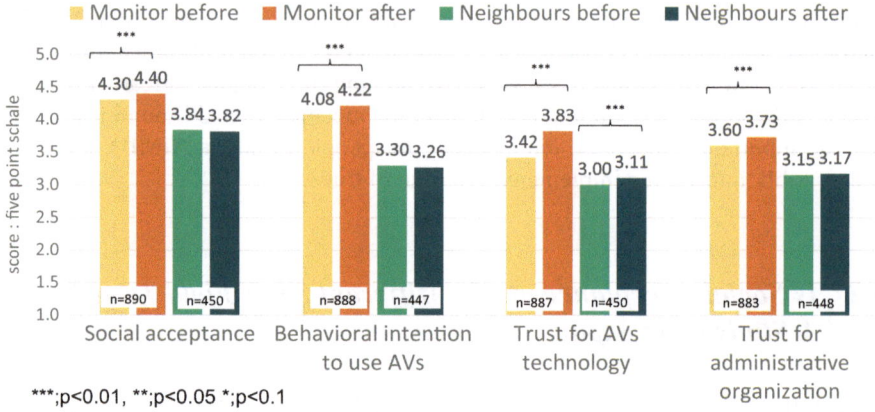

Fig. 17 Changes in attitude on the part of onboard monitors and local residents after AV field trials

constructing legal systems, insurance and other social mechanisms, and ensuring a thorough understanding of these mechanisms to increase trust in administrative agencies.

However, the field trials in 2017 were conducted for a short period of one to two weeks. In the long-term field trials in 2018–2019, which were conducted for a period of one to two months, there was no change in attitude after as compared to before the test [18]. This is thought to be because the attitude of the test participants toward AVs from the outset was high at 4.2–4.3 out of 5. In these long-term tests, the number of times that the AVs avoided vehicles parked on the street, detected weeds and planted areas (weeds and trees can change the 3D map as they grow overgrown, and this poses a major problem for the operation of AVs), stopped to yield to following vehicles and so on, or in which manual intervention by the driver was needed, were counted as incidents. The results showed that the more such incidents were encountered,

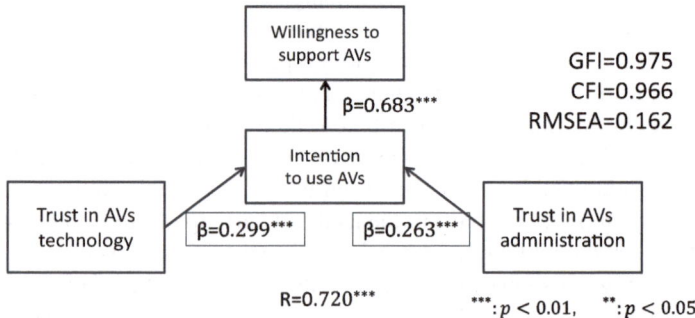

Fig. 18 Relationship of willingness to support AVs, intention to use AVs, and trust in AV administration

the more people had "near-miss" experiences, and this had a negative impact on confidence in the technology and the administration.

Based on the above, providing widespread opportunities to experience AVs through field trials has the effect of increasing positive attitude, at least among those who have a preliminary interest in participating in such tests. However, revealing the limits of the technology through incidents can adversely affect confidence, so careful planning and efforts to reduce incidents will be essential.

4 Media Coverage of AVs

This section presents the results of media analyses, particularly newspaper presentation of AVs in Japan and Germany. The media plays a role in directing the public's attention to topics related to AVs, and certain aspects of AVs, while not covering many other aspects. The results of the analyses show that the number of articles on AVs has increased significantly over recent years and identifies which topics and expectations of AVs have been shared.

4.1 The Japanese Study on AVs in Newspaper Reports

In Japan, AVs have become more familiar to the general public since the field trials on public roads began in 2017, and after driving-support technologies began to be implemented in vehicles. In addition, the discussions regarding AVs and the problems involved have become increasingly diverse. To study the social acceptance of AVs in this phase, it will be important to determine people's past and present understanding of the issues and problems relating to AVs. Even if mass media reporting may have little effect on directly changing people's attitudes, the media plays a role in setting the agenda for what issues should be discussed by selecting certain features of reality and making them more salient in the communicating text [6, 16]. Tracking the changes in the content of media reporting about AVs can help to determine what the media feels people should be discussing about AVs. This section covers a research study about media reporting in Japan [19].

The study focused on newspapers. Newspaper readership has been declining in recent years, but in many cases TV news and news reports on social networking services (such as Facebook, twitter etc.) use newspapers as a news source, so even if people do not read newspapers directly, they are exposed to newspaper articles indirectly. In addition, newspaper articles dating back to the Meiji period (1868–1912) have been archived in the form of text data, so this approach has the advantage of making it possible to conduct an analysis over a long period of time. Below is a discussion of the changes in the topics relating to AV development and introduction that have been provided to society by newspaper reporting in Japan.

4.1.1 Study Overview

First, newspaper articles on AVs were gathered from the Yomidasu Rekishikan archives of the Yomiuri Shimbun newspaper, which as of 2019 had the highest circulation. The search was conducted by searching for "automated driving + automobile" in the Yomidasu Rekishikan search engine and excluding words such as "railway". An analysis was then conducted for the total of 1,026 articles located, which were published between October 31, 1989 and December 31, 2019.

The articles were read carefully, and categorized according to the AV occurrence described in the article, the purpose of development and introduction and issues encountered, and the point of view and opinions expressed in the article. In addition, the analysis traced and considered the article content based on the historical backdrop. The topics included in the article were identified and the articles were classified into two major categories: those that dealt with the objectives of AV development, and those that dealt with problems relating to AVs. The articles were read by one author and then the classification and considerations were discussed by the co-authors [19].

4.1.2 Number of Newspaper Articles and Changes in Topics

Changes in the Number of Newspaper Articles

Figure 19 shows the changes in the number of newspaper articles relating to AVs. The first article to mention "automated driving" appeared in 1989, and between 1995 and 2005 such articles appeared as part of a discussion of Intelligent Transport Systems (ITS). In 2005, a self-driving bus was used as a means of transport at the site of Aichi Expo 2005 held in Aichi Prefecture, and as a result the number of articles increased in the years before and after this event (although system errors forced the suspension of operation prior to the conclusion of the Expo). Subsequently, between 2006 and 2012, the number of articles about AV dropped to between 0 and 4 per year. In 2009, there were zero articles for the first time in 16 years (since 1993). It is possible that this was because the news was dominated by the so-called "Lehman Shock" and the subsequent financial crisis. In August 2013, the Nissan Motor Company announced that it would begin sales of a self-driving automobile in 2020, and as a result the number of articles increased dramatically. Moreover, around the year 2016, field trials on public roads were actively conducted, and the number of articles increased.

Changes in Newspaper Articles About the Objectives of AV Development

Table 8 shows the objectives of AV development, and the categories for development objective as determined by the analysis of newspaper articles. Based on [19].

The years in which articles on the objectives of AV development first appeared, can be divided into two periods: the 1990s and the 2010s. Objectives that appeared during the 1990s include "Dreams and Romance", "Safety", "Economic stimulation",

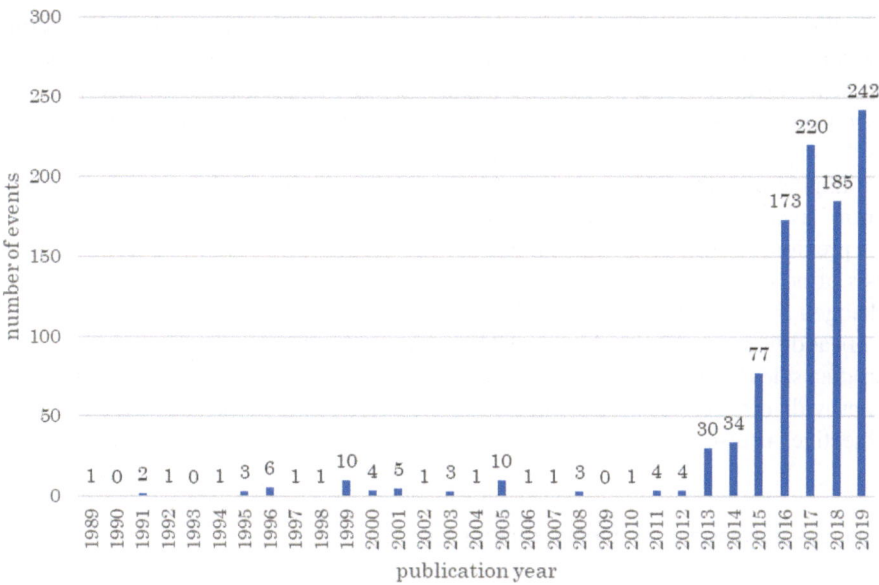

Fig. 19 Number of articles on AVs in Japanese newspapers

Table 8 Categories for objectives of AV development

Major classification	Sub-classifications (number of Articles)
Competition	International competition (135), Competition in the automobile industry (131), Competition with other industries (76)
Safety	Traffic accidents (106), Safety (61), Transit issues (1)
Economy	Economic benefits (68)
Traffic congestion	Easing congestion (30)
Going-out support	Going-out support (50)
Driver issues	Resolving the driver shortage (37), Driver burden (10), Transit in depopulated areas (9), Reducing public transport operating costs (5), Pick-up and drop-off (2)
Dreams and Romance	Dreams and Romance (21), Lifestyle (5)
Effective use of travel time	Effective use of travel time (5)
Other services	Various services (4), Vehicle dispatch (3), ATMs (1), Vending machines (1), Services in individual regions (1), Ride-sharing (2)
Protecting the natural environment	Protecting the natural environment (7)
Other	Other (25), Regional appeal (1)

"Traffic congestion", "(International) competition" etc. It appears that "Protecting the natural environment" and "Driver issues" were not recognized as major development objectives, as a period of 20 years or more elapsed between their first and second appearances. In the 2010s, new development objectives were mentioned, such as "Going-out support", and "Effective use of time." Beginning in 2015, there was a dramatic increase in the number of articles that mentioned "International competition" as an objective of AV development, and since 2013 this issue has surpassed the previously most-frequently cited issue of "Traffic safety." Judging from the analysis of newspaper articles, the major objective of AV development appears to have changed from "Traffic safety" to "Winning the international competition."

In addition, since 2016, the use of AVs to provide new services such as delivery and product sales has been much talked about, and expressions such as "AVs development to provide new services" have appeared. The background to the appearance of these objectives is discussed in detail in Sect. 4.1.3.

Changes in Reported Problems Relating to AVs

Table 9 shows the problems relating to AVs, arranged and classified in the same manner as the development objectives. Over the entire time period, the largest number of articles regarding problems concerned "development funds." With regard to the development of legal systems (laws and regulations), many articles discussed this issue, together with "Determination of responsibility". The most active discussion of "Developing international standards" occurred in 2014, and it is possible that during this period the "technical issues" relating to AVs were seen to be gradually being resolved. Moreover, it was in 2016 that a discussion first emerged regarding adverse effects, such as "relying too much on AVs", and "difficulty to change smoothly between the human driver and the system". It is possible that this occurred because field trials on public roads had begun to increase at around the same time. There was widespread detailed discussion not only of the technical problems of the individual AVs, but also of conflicts between the AVs and users. Since 2015, the number of articles has increased and a greater diversity of problems is being discussed—changes that are a point of commonality with articles about the objectives of AV development.

4.1.3   The First Appearance of "Automated Driving", and Newspaper Articles that Became a Turning Point

A qualitative analysis of characteristic articles was conducted as described below.

First Appearance of "AVs" in Newspaper Articles

The phrase "self-driving vehicle" first appeared in a Yomiuri Shimbun newspaper article in 1989. In this article, AVs were referred to as the ultimate automobile, and

Table 9 Classification of AV problems

Major classification	Sub-classifications (number of Articles)
Technology	Technology (68)
Social acceptance	Social acceptance (33)
Development of international standards	International standards (25)
Infrastructure construction	Infrastructure (18)
Funding	R&D funding (90), Financial matters (2)
Laws and insurance	Laws and regulations (70), Determination of responsibility (51), Insurance (17)
Discomfort with non-human driver	Pleasure of driving (11), Machine substituting for human driver (9)
Type of development organization	No entity to take charge domestically (8), Speeding up the decision-making process (1)
Adverse effects	Over-reliance (8), Adverse effects (5), Switching between human driver and AVs (1)
Price	Price (6), Maintenance costs (1)
Ethical issues	Ethics (5), Reason for human existence (1), Balance between convenience and safety (1)
Labour shortage	Labour shortage (5), Researcher shortage (3)
User knowledge	Education (3), Explanations (2), Licensing (2), Construction (1)
Nature of new technology	Nature of means of transport (2), New types of accidents (1)
Other	Other (20), Period of introduction (1), Time (1), Data leakage (1), Military use (1)

a vehicle exhibited at the Tokyo Motor Show was introduced as being a step closer to the dream of AVs. As can be seen from the terms "the ultimate automobile" and "dream," at that time the practical achievement of AVs was considered to be a far-off prospect.

Autonomous Vehicles as a Symbol

From November 1995 to June 2003, AV was frequently used in newspaper articles as an example of ITS technology. Nowadays, Vehicle Information and Communication System (VICS) and Electronic Toll Collection (ETC) are recognized as examples of ITS technology, but at the time, it was difficult to explain these developments in simple terms. In contrast, "automated driving" was a straightforward expression that was easy for people to picture, and presumably this is the reason that it was used as an example of ITS technology. Since 2015, AVs have been introduced as an example of the use of state-of-the-art technologies that include IoT (Internet of Things), AI, deep learning and 5G. Of the ITS technologies, VICS and ETC have been developed for practical use and have been disseminated. In contrast, AVs are still in the process of being developed, and have become an example of the application of not just automotive technology but of technologies in various other fields (communications, IT, surveying and maps, space and satellites etc.,), and as a result AVs can be considered to have become a symbol of technical development overall.

Development Objective: Preventing Traffic Accidents

The first newspaper article to express the hope that AVs could prevent traffic accidents was the Editor's Notebook that appeared in Yomiuri Shimbun on July 22, 1991. This article noted that the world's first automobile traffic accident was caused in 1769 by the world's first automobile, and it lamented that we are still plagued by traffic accidents today (1991). It then introduced the features of AVs and expressed the hope that they could reduce traffic accidents. At the same time, the article expressed doubt that an advanced unit could handle all operations instead of a human being, and expressed the view that coordination of people, vehicles and facilities would be needed to ensure road safety.

Development Objective: Economic Stimulation

Between 1996 and approximately 1999, the view of the Japanese Ministry of Construction that the economic benefits of ITS would amount to JPY 50 trillion over 20 years appeared several times in articles mentioning ITS. The mood in Japan at that time was unsettled due to the collapse of the "bubble" economy in 1992, and the Great Hanshin-Awaji Earthquake and the sarin gas attack on the Tokyo subway in 1995. Therefore, it is possible that expectations for an economic rebound due to ITS were high. In addition, twelve articles mentioned "automated driving and other ITS", implying that a climate of expectation regarding the economic benefits of AVs through ITS had been established.

Development Objective: International Competitiveness (Trauma Regarding the "Galapagos Phone" Fiasco)

The analysis of newspaper articles reveals that the development objective of international competitiveness has been mentioned since the 1990s, but it was around 2014 that a deeper discussion began. At the time, there were active efforts to formulate international standards for AVs, and nations were competing to take the lead. Japan was therefore working to take the lead in international standards formulation, and the background to this effort was indicated in the title of an article dated June 24, 2014: "Strategic move to establish a superior position for Japanese automobiles: Avoiding the "Galapagos Phone" fiasco in proposing standards for automated driving". Moreover, an article dated September 27 of the same year, entitled "IoT international standard: Learning the lessons of Galapagos-ization (analysis)", discussed the risks in these terms: "Standards wars require a great deal of time and costs, and when you are defeated, all of that effort goes right down the drain".

As noted in these articles, Japanese mobile phone manufacturers that had pursued an independent development path fell behind overseas smartphones and were knocked out of the market one after another. These "Galapagos" mobile phones did not become the international standard and were defeated in the international competition. Based on this lesson, the newspaper appears to have laid out an agenda recommending the pursuit of AV development in order to avoid going down the same path. However, there was no explanation as to why it was necessary to win the international competition. There were also many articles that limited themselves to saying that there was intense international competition, something that might lead to confusing means and objectives, or turning the means into an objective.

Development Objective: Support for Senior Citizens and Vulnerable Transport Users

Since around 2015, articles about the use of AVs to assist senior citizen mobility have appeared. At that time, accidents involving elderly drivers and the increase in the fatalities of elderly drivers in traffic accidents had become an issue. In some cases, going-out support for senior citizens was talked about in the same way as supporting disabled persons, indicating expectations for AVs as a means of providing freedom of mobility to people without being dependent on their physical abilities.

Problem: Accidents Caused by Over-Reliance on AVs

In April 2017, an article entitled "If you trust 'refraining from braking'—Self-driving vehicle drives into people without stopping" appeared. In November of 2016, a person riding in a test vehicle was prompted to use the braking assist feature, but the system failed to function and the vehicle plowed into people, resulting in injuries. The article cited this incident and also introduced a notification from the police in April 2017 about this incident. The article said that, in some cases, the commercially-available AD features do not function properly due to weather or ambient conditions, but sales personnel and drivers have used these features without understanding their limitations, with the result that such incidents have occurred. In all, eight articles sounded the alarm about the fact that such incidents have been increasing each year, and that drivers must not place too much reliance on automated driving features. As vehicles equipped with AD features become more and more popular, finding a way to prevent drivers from excessive reliance on AVs is seen as a problem that needs to be resolved.

Problem: Determination of Responsibility for Accidents and Development
of Legal Systems

The first newspaper article to contain a discussion of the determination of responsibility for traffic accidents caused by AVs appeared in November 2000. The article cited this as one problem that needed to be resolved, but it did not discuss the issue in detail. The next such article appeared in 2003 and pointed out that, in the event that an accident occurred while the AVs were in operation, the manufacturer may have to assume responsibility, and as a result companies were not enthusiastic about pursuing development. Of the 51 articles that discuss the determination of responsibility for accidents, 20 limit themselves to merely pointing out the existence of the problem of determining responsibility, while seven suggest the possibility that manufacturers may be held responsible. In contrast, only a single article deals with the operator's responsibility, and a single article touches on the possibility that there will be no-one to be held responsible. In the discussion of the determination of responsibility for an accident, the need to develop legal systems has been pointed out (7 articles). There were also articles mentioning the existence of unease caused by the fact that determining responsibility is unclear, and pointing out that public opinion should be reflected in the creation of rules relating to this issue.

Problem: Development Funding

From 1994 to 2005, there were occasional articles about the national government securing a budget to support AV development as a part of ITS. Since 2014, however, these articles primarily pointed out that development expenditure by private companies had increased. Beginning in 2016, group company reorganizations and capital alliances were conducted, and articles noted efforts to reduce the associated increase in AV development expenditure through joint research.

The analysis of articles seems to show that neither the national government nor private sector companies can withdraw from AV development, due to the fact that companies must recover research and development expenses through profits over the course of 30 years. AVs are a symbol of technical development in a variety of industries, such as the information industry, aerial survey industry and so on, and that international competition is heating up. As of 2021, it is unclear when fully-fledged AV implementation will become a reality.

Diversification of Problems

As in the case of AV development objectives, the problems with AVs that have been pointed out have become diverse in recent years. One reason for this is likely the increase in industries involved in AV development and the increase in elemental technologies. The diversification of problems can also be thought of as the result of active field trials being conducted on public roads, providing increased opportunities for people to experience AV systems. These systems are in full-fledged use in Sakaimachi, Ibaraki Prefecture, and driving safety-support technology is being incorporated into commercially-available vehicles. All of these developments have led people to feel that AVs are familiar to them, resulting in increased discussions about AVs.

By tracing the changes in AV development objectives and problems through the analysis of newspaper articles, it was possible to clarify the process by which AVs became symbolic of the sum of various technologies, as well as changes to shifting development objectives, concerns that means have become objectives, diversification of problems and so on. In the future, it is likely that situations that cannot be anticipated at this stage will become problems. In order to implement AVs in society, it will be essential to anticipate unforeseen situations and expect the unexpected, and to be prepared to deal with such situations in a forthright manner.

4.2 A Brief Analysis of Media Reporting on AVs in a German Newspaper

AV is associated in the German scientific, public and political discussion with often very different expectations and promises of solutions. In many cases, safety issues are in the foreground. Some see AV as the key enabler for a transition towards sustainable mobility, while others focus on the international competitiveness of the German automotive industry. Against this background, it is interesting to examine which expectations are placed in the foreground in media reporting. Within the framework of the CADIA project in Germany (for CADIA see the introduction), a small media analysis was carried out. The results are summarised below. The analysis concentrated on "Handelsblatt" ("Commerce paper"), a German daily business newspaper with a strong focus on economic and financial topics. The circulation is 127,280 copies, and a limited range of the reporting is accessible online. In 2020, Handelsblatt was the most cited business newspaper in Germany. Since Handelsblatt repeatedly deals with the economic potential of new technologies, a relatively high presence of the AV topic could be expected.

As search strings, the German versions of the following terms were used: "automated driving", "autonomous driving", "automated cars, autonomous cars".

The analysis focused on the number of articles (661 overall) as well as the expectations or promises related to AVs and/or future developments. In the scientific and public discourse on AVs, a variety of expectations is usually reported. For the analysis, we differentiated between the following expectations:

1. AVs as a means to ensure international competitiveness of German industry
2. AVs as means to improve safety in traffic
3. AVs as an enabler for a sustainable transition in the mobility sector
4. AVs enable individual mobility for disabled persons and children/teenagers

5. AVs enable an increase in the efficiency of the mobility system (optimization of traffic flows on roads and at intersections)
6. AVs as an enabler for travel time savings (reading, working etc. while travelling)
7. AVs only briefly mentioned without clear expectations or promises.

Results are plotted in Fig. 20. Until 2018, the amount of articles increased steadily. It can be assumed that the decline in 2020 is at least partly due to the dominance of COVID-19 in public discourse and a corresponding shift in the focus of media reporting.

Casual mentions in particular fell sharply in 2020 and 2021. The expectation "international competitiveness" was relevant in 45% of the articles. It dominates the sample very strongly, with approximately half of all articles primarily dedicated to this subject. Given the orientation of Handelsblatt, a focus on economic issues was to be expected. Nonetheless, the overall result is in line with the observations of Jelinski et al. (2021), who performed a standardized content analysis of five German media outlets between 1 May 2017 and 31 October 2018, and found a similar dominance of economic reporting. The topic of "safety", which was important in 88 articles, follows at a considerable distance. The other categories are rather rare, although the topic "sustainable mobility" was relevant in 23 articles in connection with AV.

Overall, it is evident that AV is strongly framed as an economic issue, which is primarily expected to lead to safety benefits. Other socially relevant aspects, such

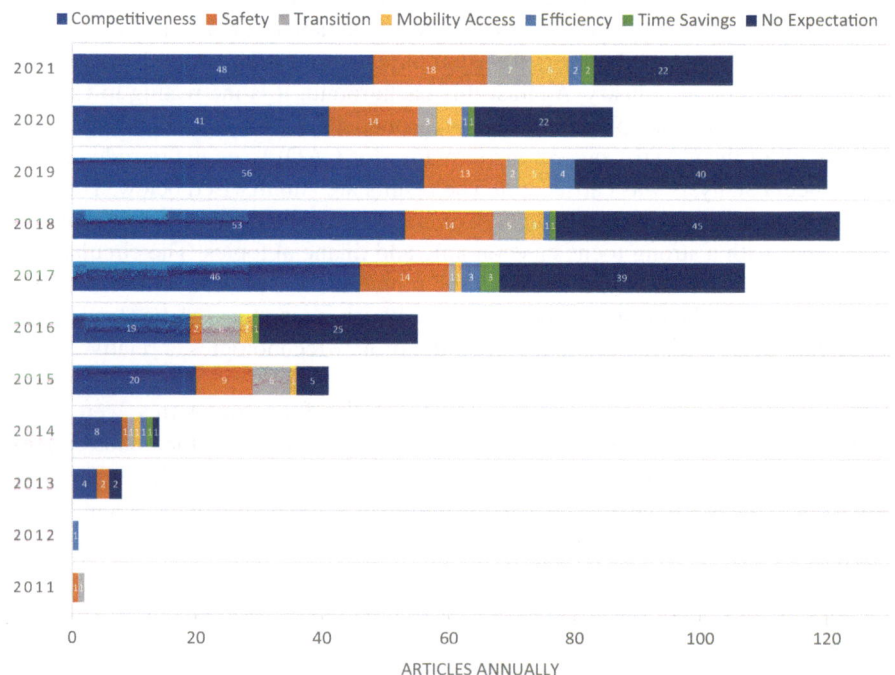

Fig. 20 Results of the media analysis of the German newspaper "Handelsblatt"

as sustainability, mobility enablement or the use or perception of travel times, are addressed only to a limited extent. The reporting therefore focuses primarily on AV as product and less on its social embedding.

4.3   Conclusions of Media Analysis

The media analyses carried out in Japan and Germany differ significantly as regards the general methodological approach, the selected publications, the time span covered and the way the articles were classified. Nevertheless, some interesting commonalities can be observed. There is a strong increase in the number of articles starting between 2013 and 2015. In Japan, this take-off phase starts somewhat earlier than in Germany. The strong emphasis on economic aspects can be demonstrated in both countries. Above all, the international competitiveness of the respective industries is clearly in the foreground. The more detailed analysis of the Japanese articles further shows that in Japan issues of regulation, legal settings and related questions of responsibility were mentioned quite frequently. In both cases, the direct traffic effects (with the exception of safety), as well as environmental impacts and social aspects of AVs, are clearly less in focus. The potential of AVs to trigger a transition of the mobility system with far-reaching changes for society, which has been widely discussed in the scientific community, at least in Germany, is hardly reflected in the media analyses conducted in the two countries.

5   Conclusion and Avenues for Further Research

This chapter first introduced various debates on social acceptance. We presented the broad definition of social acceptance of a technology, by illuminating different dimensions of acceptance—acceptance subjects and acceptance objects and the relationships between them, and suggested encompassing various classifications in each dimension. The various studies introduced following the conceptual discussions focused on different acceptance phenomena and the contributing elements that influence them. The results reveal various commonalities but also differences between Germany and Japan.

In all surveys that the authors conducted in both countries, they found the tendency that respondents in Japan offer less extreme answers and lie more towards the centre compared to the German respondents. A matter of future research will be whether this tendency will be observed in other surveys, regardless of the topic. We think that some of the differences in response behaviour can be attributed to different cultural settings. For example, the results of our explorative questionnaire survey in 2020 show that German respondents are relatively positive when it comes to reducing CO_2 emissions with AVs. The high importance of environmental issues in German society certainly plays a role here. To give another example from the same survey: When asked whether road safety regulations should be relaxed to make AVs easier to implement, the respondents in Japan were surprisingly positive. This may be related to the fact that "deregulation" is generally framed quite positively in Japanese discourse. The larger-scale population survey in 2021 revealed interesting and sometimes surprising findings. Respondents in Japan are more optimistic about the impact of AVs on cost reduction of mobility services, whereas German respondents are more optimistic about impacts of AVs on stationary traffic and traffic flow. There are some highly interesting similarities with regard to the expected long-term effects in both countries. The statistical analysis reveals that most current expectations about the long-term effects of AVs may be shaped by one potential factor. Future research should identify this latent factor and compare it in both countries. Another interesting finding is that, in both countries, regulatory measures to improve safety are most welcomed, while measures that aim at relaxation of regulatory frameworks are most opposed. An intriguing difference can be seen in preference for a functional design of AD. Two-thirds of the German respondents would like to be able to intervene in AVs if this is perceived as necessary to avoid an accident. Interestingly, less than half of the respondents in Japan share this view. As regards the set of questions about wellbeing in different use cases, it is of considerable interest that the patterns of answers are quite similar in both countries, even though respondents in Japan tend more towards the middle of the options. What is striking is that in both countries, wellbeing in new mobility services is not only determined by the degree or concept of automation, but also by other design and service factors. It would be important to dig deeper into this finding in follow-up projects to get a better understanding of these "other" design and service factors.

The results of the survey conducted in 2017 where field trials have taken place in Japan suggest that participation in field trials can enhance social acceptance, as measured by the increase in trust in the technology and the administrative and business actors involved, the intention to use AD public transport, and the willingness to support society with public transport that uses AD. However, in particular in Japan, so-called NIMBY-effects might reduce the willingness of affected citizens to accept field trials. At the same time, it is striking that many more German respondents than those in Japan disagree with the idea of creating a *society realizing an autonomous vehicles system.*" The news article analyses show that, in both countries, increasing (or sustaining) the international competitiveness of the national industry has been the central topic of AD in the media during recent years.

Without a doubt, AVs are a technology with very great potential for both countries in terms of transport developments, but also in terms of economic and social developments. It is therefore very important to understand how different technical and non-technical factors, including social acceptance, influence the development of AVs and where there is leeway for political influence. In order to achieve a "soft landing" for AVs in society, discussions that involve government, academia and the private sector will be essential, and comparisons with countries that have different societal and cultural backgrounds will also be important to contribute to the discussion. The results presented in this chapter show that a comparative perspective can provide important insights. It would now be very important to build on these findings and to provide appropriate scientific comparative support for the further development of AVs in the two countries. There is much to suggest that AVs will develop dynamically over the next few years, and that this period will set the course for many future development directions. It is hoped that cooperation between Japan and Germany at both the governmental and research level in a variety of venues can be enhanced.

References

1. BMVI-Bundesministerium für Verkehr und digitale Infrastruktur (2017). Bericht zum Stand der Umsetzung der Strategie automatisiertes und vernetztes Fahren. Berlin, November 2017.
2. Becker, F., & Axhausen, K.W. (2017). Literature review on surveys investigating the acceptance of automated vehicles. *Transportation*, 44, 1293–1306.
3. Becker, S., & Renn, O. (2019). Akzeptanzbedingungen politischer Maßnahmen für die Verkehrswende: Das Fallbeispiel Berliner Mobilitätsgesetz. In Fraune, C., Knodt, M., Gölz, S. Langer, K. (Eds.), *Akzeptanz und politische Partizipation in der Energietransformation* (pp. 109–130), Springer Fachmedien: Wiesbaden.
4. Bundesregierung (2019). Aktionsplan Forschung für autonomes Fahren. Ein übergreifender Forschungsrahmen von BMBF, BMWi und BMVI. Berlin, July 2019.
5. Burningham, K., Barnett, J., & Thrush, D. (2006). The limitations of the NIMBY concept for understanding public engagement with renewable energy technologies: a literature review, published by the School of Environment and Development, University of Manchester, Oxford Road, Manchester M13 9PL, UK.
6. Entman, R. M. (1993). Framing: Toward Clarification of a Fractured Paradigm. *Journal of Communication*, 43(4), 51–58. https://doi.org/10.1111/j.1460-2466.1993.tb01304.x.
7. Fleischer, T., Schippl, J., Yamasaki, Y., Taniguchi, A., Nakao, S., & Tanaka, K. (2020). Social Acceptance of Automated Driving in Germany and Japan-Conceptual Issues and Empirical Insights, SIP-adus (Innovation of Automated Driving for Universal Service) Workshop 2020, held Online, 9th November, 2020.
8. Fleischer, T., Schippl, J., Yamasaki, Y., & Taniguchi, A. (2021). Social Acceptance of Automated Driving: Some Insights from Comparative Research in Japan and Germany. Proceedings of the 27th ITS World Congress, Hamburg, 11.-15.10.2021.
9. Fraedrich, E., & Lenz, B. (2016). Societal and Individual Acceptance of Autonomous Driving. In Markus Maurer, J. Christian Gerdes, Barbara Lenz and Hermann Winner (Eds.), *Autonomous Driving. Technical, Legal and Social Aspects* (pp. 621–640). Berlin, Heidelberg: Springer Berlin Heidelberg.
10. Gkartzonikas, C., & Gkritza, K. (2019). What have we learned? A review of stated preference and choice studies on autonomous vehicles. *Transportation Research Part C: Emerging Technologies*, 98, 323–337. https://doi.org/10.1016/j.trc.2018.12.003.
11. Grunwald, A. (2005). Zur Rolle von Akzeptanz und Akzeptabilität von Technik bei der Bewältigung von Technikkonflikten. *TATuP—Zeitschrift für Technikfolgenabschätzung in Theorie und Praxis, 14*(3), 54–60. https://doi.org/10.14512/tatup.14.3.54
12. IT Headquarters. (2019) Public-Private ITS Initiative/Roadmaps 2019. https://japan.kantei.go.jp/policy/it/2019/2019_roadmaps.pdf.
13. Itsubo, S., Tamada, K., Sawai, S., & Taniguchi, A. (2018). Analysis of Societal Acceptance of Proving Tests for Autonomous Vehicle Services Based at Michi-no-Eki, etc. *Proceedings of Infrastructure Planning* (CD-ROM), Vol. 57.
14. Kawashima, Y., Taniguchi, A., Itsubo, S., Tamada, K., & Sawai, S. (2018). Determining Factors for Societal Acceptance of Autonomous Vehicle Public Transport Services. *Proceedings of Infrastructure Planning* (CD-ROM), Vol. 57.

15. Lucke, D. (1995). Akzeptanz. Legitimität in der „Abstimmungsgesellschaft". *Leske + Budrich*: Opladen.
16. McCombs, M. E. (1992). Explorers and surveyors: Expanding strategies for agenda-setting research. *Journalism & Mass Communication Quarterly*, 69(4), 813–824. https://doi.org/10.1177/107769909206900402.
17. Meyer, U. (2016). Innovationspfade. Evolution und Institutionalisierung komplexer Technologie. *Springer VS*, Wiesbaden.
18. Minamite, K., Taniguchi, A., Itsubo, S., & Kawashima, Y. (2020). A psychological process model of objective incidents and perceptions of approval or disapproval in field operational tests using autonomous vehicles. *62nd Proceedings of Infrastructure Planning* (CD-ROM).
19. Miyadai, K., Tanaka, K., Nakao, S., & Taniguchi, A. (2020). Qualitative Analysis of Newspaper Reporting on Changes in Objectives for Introduction of Automated Driving Systems. *62nd Proceedings of Infrastructure Planning* (CD-ROM).
20. Nakao, S., Tanaka, K., Taniguchi, A., Kanzaki, N., Kukita, M., Miyadai, K., & Minamite, K. (2020). Comparative Analysis of Japan, the U.K. and Germany with regard to Social Acceptance of Automated Driving Systems, Focusing on Tones relating to AVs. *62nd Proceedings of Infrastructure Planning* (CD-ROM).
21. Nastjuk, I., Herrenkind, B., Marrone, M., Brendel, A. B., & Kolbe, L. M. (2020). What drives the acceptance of autonomous driving? An investigation of acceptance factors from an end-user's perspective. *Technological Forecasting and Social Change, 161*, 120319.
22. Pratt, G. (2017, December 20). *How Toyota wants to reduce the 1.3 million lives lost a year in traffic accidents*. Toyota Canada Newsroom. http://media.toyota.ca/releases/how-toyota-wants-to-reduce-the-1-3-million-lives-lost-a-year-in-traffic-accidents.
23. Reichenbach, M. & Fleischer, T. (2022). Zwischen Ambition und Umsetzung: Institutionalisierungs-prozesse als Kernherausforderung der Mobilitätswende?. In: D. Sack, H. Straßheim, K. Zimmermann (Eds.), *Renaissance der Verkehrspolitik. Politik- und mobilitätswissenschaftliche Perspektiven*. Springer VS.
24. Rogers, E. M. (2003). Diffusion of innovations (5th Edition). New York: Free Press.
25. Scott, W.R. (2014): Institutions and Organizations: Ideas, Interests, and Identities. 4th Edition. *SAGE*, Los Angeles.
26. Shakaiteki Juyōsei. (n.d.). In Digital Daijisen. Retrieved July 8, 2022, from https://kotobank.jp/word/%E7%A4%BE%E4%BC%9A%E7%9A%84%E5%8F%97%E5%AE%B9%E6%80%A7-681590.
27. Sheller, M., & Urry, J. (2016). Mobilizing the new mobilities paradigm. *Applied Mobilities, 1*(1), 10-25. https://doi.org/10.1080/23800127.2016.1151216.
28. Tanaka, K., Nakao, S., Taniguchi, A., Kanzaki, N., Kukita, M., Miyadai, K., & Minamite, K. (2020). Comparative Analysis of NIMBY Sentiment in Japan, the U.K. and Germany with regard to Social Acceptance of Automated Driving Technology. *62nd Proceedings of Infrastructure Planning* (CD-ROM).
29. Taniguchi, A. (2019). Societal Acceptance of Automated Driving: Determining Factors and the Potential for Change. *JSAE Journal* (Journal of the Society of Automotive Engineers of Japan, Inc.: Special Feature "Automated Driving and Societal Acceptance"). 73(2), 44–50.
30. Taniguchi, A., Tomio, Y., Kawashima, Y., Enoch, M., Ieromonachou, P., & Morikawa, T. (2017). Societal Acceptance of Autonomous Vehicle Systems: Willingness to Support AVs and Risk Perception. *Proceedings of Infrastructure Planning* (CD-ROM), Vol. 56.
31. Taniguchi, A. (2020). A Comparison of Social Acceptance of Autonomous Vehicles At Two Time Points in Japan, UK and Germany. *Proceedings of Infrastructure Planning* (CD-ROM) (Japan Society of Civil Engineering), Vol. 62.

Transportation Effects of Connected and Automated Driving in Germany

Michael Schrömbges, Dennis Seibert, and Nina Thomsen

Abstract Connected and automated driving (CAD) is likely to affect the German transportation system. Three consecutive models assess the effects of private automated vehicles (PAV) and shared automated vehicles (SAV) on car ownership, car stock, and travel demand in 2050 based on different scenarios. Firstly, a car ownership model (COM) estimates car availability at household level including changes in accessibility through CAD. Based on the year of market entry and additional costs, a car stock model (CAST) quantifies the diffusion of CAD within the German car fleet in 2050. Finally, the effects of CAD on transportation volumes and key indicators for SAV services are determined using the national travel demand model of Germany (DEMO). The model results for PAV show an increase in ownership by up to 1% with a 44% diffusion of Level 4+ cars in 2050. They account for over 50% of kilometers driven and increase overall vehicle kilometers traveled by 3%. On the contrary, SAV will reduce car ownership in urban regions by up to 4%, but increase vehicle kilometers traveled by 5%. Automation improves the transportation system and makes traveling easier. But to cope with the environmental implications, it is necessary to provide a political framework which stresses the advantages of CAD.

M. Schrömbges (✉)
Institute of Transport Planning, RWTH Aachen University, Aachen, Germany
e-mail: schroembges@isb.rwth-aachen.de

D. Seibert · N. Thomsen
Institute of Transport Research, German Aerospace Center (DLR), Berlin, Germany
e-mail: d.seibert@oeko.de

N. Thomsen
e-mail: nina.thomsen@dlr.de

© The Author(s) 2025
C. Eisenmann et al. (eds.), *Acceptance and Diffusion of Connected and Automated Driving in Japan and Germany*, https://doi.org/10.1007/978-3-031-59876-0_6

1 Introduction

The transportation system experiences disruptive changes through technological development, such as electrification and digitalization. One highly researched and current example is connected and automated driving (CAD), which will likely transform the transportation system. CAD consists of two aspects: the first part, "connected", means vehicles communicate with each other (vehicle-to-vehicle or V2V), or with infrastructure such as traffic lights (vehicle-to-infrastructure or V2I) [20]. This function is the basis for the second part, "automated", as it provides the technical requirements along with the onboard sensors. Therefore, both terms in combination lead to "connected and automated driving".

CAD refers to vehicles driving themselves without manual intervention, in specified or all situations. According to the definition of automated driving by the Society of Automotive Engineers SAE International [28], CAD refers to Level 4 and 5 (for levels of CAD see the Introduction), throughout the remainder of this chapter labeled as Level 4+. This includes private automated vehicles (PAV) and the emergence of new transportation modes with shared automated vehicles (SAV). The advantages of CAD are expected to be manifold, including:

- ensuring mobility for people with reduced mobility [41, 44, 48]
- maintaining accessibility in regions with low demand [25, 41]
- improving the efficiency of the transportation system [6, 48]
- increased comfort and the possibility to use the travel time alternatively [6, 25, 41]
- reduction of transportation-related emissions [41, 48].

Nevertheless, the expected effects only occur if the corresponding regulatory framework exists and society accepts the technology. The chapter "Governance, Policy and Regulation in the Field of Automated Driving: A Focus on Japan and Germany" of this book analyses the regulatory framework in Japan contrasted with Germany, while the chapter "Social Acceptance of CAD in Japan and Germany: Conceptual Issues and Empirical Insights" offers comprehensive insights on the social acceptance of CAD in Germany and Japan. As CAD is not yet part of the transportation systems, extensive research analyses and simulations are being conducted for different future scenarios to pave the way for transforming the transportation sector. With the technology and regulatory framework developing rapidly, the analyses at hand use a future scenario for the year 2050 in which automated vehicles such as PAV and SAV are available on the market. To quantify the impact of CAD on the transportation system, a framework consisting of three models is developed. The first is a car ownership model (COM) which forecasts the effect of CAD on private car ownership in households. The second is a car stock model (CAST) which simulates the diffusion of automated vehicles in the car stock until 2050. The third is a travel demand model (DEMO), which models changes in the transportation system. This combined model set provides a powerful framework for detailed quantification of the transformation effects of CAD in Germany.

The chapter begins with an overview of research regarding modeling the effects of automation. This is followed by a section about the modeled scenarios for CAD and the model framework. After providing the results of the model, a discussion of its limitations and a summary of the results concludes the chapter.

2 Literature Review

The transformation to a future sustainable transportation system requires suitable strategies and policies. A prerequisite for developing such guidance is understanding the relationships and dependencies in the mobility sector. In this context, models in the field of transportation research help to identify and quantify these mechanisms. Thus, the analysis of the effects of CAD on the private car market and new services needs an efficient modeling framework [61]. One crucial part of this framework models changes in car ownership, car stock, and travel demand. Although models of local or regional scope allow simulation of complex dependencies due to data availability and computational power, using models of a national scope has the advantage that the evaluation of transportation-related effects on a nation as well as differentiation for urban and rural regions are possible. Thus, policy recommendations can be derived at the highest level. It follows that the development of a dedicated model framework to assess the effects of automation on a national scale can be a crucial tool, which can also provide references and frameworks for simulations in specific regions.

Simulating the effects of CAD, beginning with car ownership, through car stock up to travel demand, is a dedicated and complex task. It covers long-term decisions about car ownership and type of automation, and short-term decisions such as mode choice [34]. Due to the wide range of decisions, it is worthwhile to analyze each model category alone, before evaluating comprehensive models.

Aspects affecting conventional car ownership are well researched [2, 3, 30, 31, 33]. The emergence of automation augments these factors through acceptance and adoption of new technologies (see chapter "Social Acceptance of CAD in Japan and Germany: Conceptual Issues and Empirical Insights", or e.g., Acheampong and Cugurullo [1]; Fraedrich and Lenz [16]; Hossain and Fatmi [26]; Lavieri et al. [40]), willingness to pay for automated cars and services (see chapter "Business Analysis and Prognosis Regarding the Shared Autonomous Vehicle Market in Germany", or e.g., Daziano et al. [10]; Gkartzonikas and Gkritza [21]), and changes in accessibility because of travel time savings, as the travel time can be used differently [35, 36, 52]. As automated vehicles are currently not available on a large scale, it is not yet possible to measure the impact of CAD on car ownership. Instead, research sheds light on the effects of automation by conducting surveys about (likely) behavior, or simulating automated systems.

This chapter focuses on modeling the implications of CAD on car ownership due to the additional accessibility caused by travel time savings. These savings occur because people can use the travel time otherwise, so the value of travel time reduces.

There are only a few models and studies that evaluate the impact of automation on car ownership [19]. Some of these models optimize vehicle usage by replacing private cars, either with PAV to share within households, or with SAV.

Schoettle and Sivak [47] use a National Household Travel Survey (NHTS) to identify idle times of household cars, in which other members of the household could use the car. By optimizing the time window for vehicle trips, they found that car ownership might reduce by up to 43%. As the sample does not include origin and destination locations, the model does not account for travel times between household members located at different places. Therefore, the estimation provides an upper boundary for car-ownership reduction. Zhang et al. [60] extend the previous methodology by adding travel patterns to a travel survey. They show that optimizing vehicle trips between household members can reduce car ownership by almost 10%. Further research simulates the effect of SAV fleets on car ownership, when people would be willing to give up their private cars. Results indicate that each SAV could replace between 10–12 private vehicles [7, 14, 29]. On the contrary, Stinson et al. [53] observe an increase of up to 6% in vehicle ownership. On the basis of a survey with over 600 participants, they perform a choice modeling of vehicle ownership in the context of a sharing economy. In the simulated scenarios, private vehicles generate income during idle times as they serve trip requests from strangers, such that owning a car becomes more affordable. As the previous studies show, vehicle ownership might increase or decline depending on the pathway of automation. Gruel and Stanford [22] describe these tendencies in their conceptual study. As private automation makes traveling by car more attractive it is likely to increase vehicle ownership. But as vehicle-sharing services make alternative modes more attractive, vehicle ownership might decline.

In addition to car ownership, the development of the car stock is crucial for simulating the effects of CAD. Therefore, car stock models forecast the diffusion of new and used cars as well as the existence of different levels of automation. Studies that model the dynamic diffusion of PAV in the passenger car stock are rare. Stock shares of PAV are mostly assumed rather than modeled. To assess the impact of CAD, some studies assume scenarios with a fixed PAV diffusion and compare different diffusion levels, for example Sonnleitner et al. [50]. However, full automation of the passenger car stock is unlikely in the foreseeable future for two reasons. First, additional costs of the technology compared to existing conventional vehicles and possible reservations of car purchasers are likely to make market shares of automated vehicles increase only gradually over time. Second, changes to the passenger car stock happen very slowly due to the long duration that passenger cars remain in the stock until they are scrapped or exported. Hence, even if PAV reach a significant market share within the next decade, the passenger car stock will likely remain a mix of PAV and conventional vehicles for another two decades after that.

The changes in car ownership and different levels of car automation in the car stock affect travel behavior. Car ownership determines the availability of a car and thus, the possibility to travel by car, whereas automated cars allow people to travel more safely and comfortably. Thus, in the context of CAD, changes in the destination choice, the mode choice and the route choice are likely. Travel demand models evaluate such changes in travel behavior. Soteropoulos et al. [51] provide an overview of current research findings regarding the impact of automated vehicles on travel behavior. The focus of past research has often been the implementation in travel demand models and traffic simulations, deriving effects on network capacity or performance of vehicle sharing systems. Many studies have been conducted using agent-based models and focused on the implementation of demand-responsive transportation with fixed demand (see Bischoff and Maciejewski [7]; Fagnant and Kockelman [15], or Richter et al. [46]). Regarding a national scale, microscopic simulations are not applicable with current computation limitations, and macroscopic travel demand models are preferred. A successful implementation on a macroscopic scale is Friedrich et al. [18], where a modeling framework to depict automated driving in macroscopic travel demand models is introduced, using findings from microscopic simulations and practice tests to model the effects on road capacity. They also describe methods to model demand and to implement ride-sharing services. These findings are applied in Sonnleitner et al. [50], where different diffusion scenarios with mixed traffic are analyzed for the city of Stuttgart and its surroundings. There is, however, no separate model step which includes the impact on car ownership and diffusion of PAV, as it is modeled in gradually increasing fixed diffusion rates. The analysis of effects is reflected in traffic performance, as there are different driving modes for driverless vehicles and demand calculation. This framework provides a straightforward approach for the inclusion of automated driving in macroscopic models. However, there are differences between a regional model and modeling on a national scale, which require alterations in methods.

Modeling the effects of CAD on car ownership, car stock, and travel demand separately provides good insight into future developments. By combining models, new relationships between variables occur, resulting in additional and detailed evaluation possibilities. Moreover, the accuracy of the model output increases as modeled predictions replace general assumptions, e.g., a constant share of automated vehicles changes to simulated variable rates. Despite the extra understanding of CAD implications, these model sets are rare because of the complex interaction between the models. Kröger et al. [38] provide an example where a vehicle technology diffusion model is combined with an aggregated travel demand model to assess diffusion rates of PAV and the impact on vehicle mileage and modal splits in Germany and the United States of America (USA). PAV diffusion is modeled based on observed market intake of automated cruise control systems in the different vehicle segments. The demand model is calibrated using NHTS, and PAV are introduced with a lower perceived travel time. This approach is further applied in Kröger et al. [39], who provide a nationwide assessment of the effects of automation, including both PAV and SAV for different scenarios in Germany. However, spatial effects can only be assumed as there is no network distribution. A classic macroscopic travel demand model with an upstream model for car ownership and PAV diffusion would provide a road network and cell structure, which opens up more possibilities regarding effects on road capacity, travel volumes and travel times. Llorca et al. [43] simulate the impact of PAV on household relocation in the Munich metropolitan area of Germany. They use a combination of an agent-based land-use and a transportation model. The land-use model contains a synthetic population including car ownership. Based on household or person attribute changes, car ownership diffuses over time. In addition to car diffusion, the households choose between conventional and automated vehicles as a function between household income and cost ratio between automated and conventional cars. As the study focuses on household relocation, there are currently no scenarios dealing with shared vehicle services. However, further attributes of the built environment such as available parking places are included. All information enters the travel demand model resulting in modal share for commute trips of approximately 90% by PAV and up to a tripling of vehicle distance traveled.

The study at hand contributes to the modeling of CAD implications for the transportation system by providing a dedicated nationwide combined simulation of car ownership, car stock, and travel demand.

3   Methodology

In order to assess the effects of CAD on the transportation system, different scenarios represent possible pathways for future automation. These scenarios are simulated within the model framework, providing insight into the distinct strategies and regulatory frameworks, such as possible services for integration of CAD into public transit, or definition of service areas required for sustainable transport systems.

3.1   Pathways for the Implementation of Automated Vehicles

As CAD is still under development, the future layout of automated vehicles within the transportation system is uncertain. Concerning technology readiness, it is widely debated when PAV or SAV will enter the market. As for fully automated vehicles, the predictions range from the end of the 2020s until the 2060s [42, 55]. Some experts are even skeptical about fully automated vehicles in general [45]. Besides technological hindrances, there might be further limitations due to cost efficiency, user acceptance, and policy regulations. The cost efficiency of ride-hailing services is discussed in the chapter "Business Analysis and Prognosis Regarding the Shared Autonomous Vehicle Market in Germany", user acceptance in the chapter "Social Acceptance of CAD in Japan and Germany: Conceptual Issues and Empirical Insights", and policy regulations in the chapter "Governance, Policy and Regulation in the Field of Automated Driving: A Focus on Japan and Germany".

This research focuses on the maximal effects of CAD as an upper boundary of implications on the transportation system, regardless of the previously-mentioned restrictions. Thus, the reference year for the models is the year 2050, as fully automated vehicles will likely exist in the mass market by then [42]. Furthermore, to deal with the uncertainties of CAD implementation and to draw conclusions about different pathways and regulations, the scenarios differ in two points:

- The availability of PAV and the design of SAV service varies from no automation (*scenario (0)*) up to full automation (*scenario (3)*). In order to distinguish the effects between PAV and SAV, one scenario only allows for PAV (*scenario (1)*). Moreover, SAV may operate either as door-to-door taxis (*scenario (2a)*) or as a feeder to public transit (*scenario (2b)*). To analyze the influence of the operation design, each service option is simulated separately. Finally, the scenario of full automation (*scenario (3)*) allows for PAV and both SAV services of door-to-door and feeder jointly.
- The spatial context divides Germany into urban and rural regions. For simplification and model runtime, municipalities with more than 100,000 inhabitants are classified as "urban" and all other regions as "rural". In addition, Germany is classified into four different region types [9] to increase the accuracy of the estimations (see Fig. 1). SAV are likely to operate in limited areas defined by the operator, as is the case with existing public transit. Therefore, one scenario reflects SAV operating in urban regions as a door-to-door service (*scenario (2a)*), as the high demand is likely to result in viable economic conditions. This assumption is further suggested by various field tests in urban regions in Germany [56, 57] and worldwide (e.g., Cruise, Waymo, Apollo Go). As SAV might reduce operational costs for public transit in areas with low demand due to reduced need for personnel, another scenario shows the application of such services as feeders in rural regions (*scenario (2b)*).

The combinations of the different variables result in a set of five scenarios (Table 1).

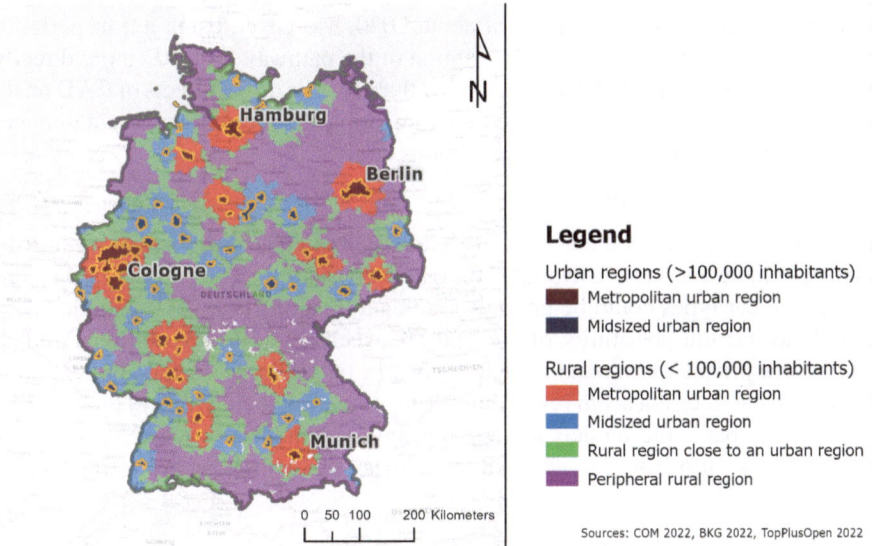

Fig. 1 Regional classification in urban and rural regions by a four-class typification [8, 9]

Table 1 Overview of CAD scenarios

	Scenario name	Private cars	Mobility as a service
(0)	Reference 2050	No automation	Not existent
(1)	Private car automation	Fully automated cars available	Not existent
(2a)	SAV in urban regions	Fully automated cars available	Door-to-door service in urban regions (more than 100,000 inhabitants)
(2b)	SAV in rural regions	Fully automated cars available	Feeder to railways in rural regions (less than 100,000 inhabitants)
(3)	Full automation	Fully automated cars available	Fully existent

(0) *Reference* 2050

The reference scenario (*scenario* (0)) represents the demographic situation of the year 2050, based on the population projections of the EU [12], Germany [11], and the Federal Institute for Research on Building, Urban Affairs and Spatial Development [5]. The projections assume that the German population will shrink by 0.6% to approx. 82.7 million in 2050. There will be approximately 22.5 million persons aged above 65 years, an increase of over 24%, showing an aging population. Concerning urban and rural population movement, the projection expects 37.49 million (+2.1%) persons living in metropolitan urban regions (red), 15.98 million (−2.5%) in midsized urban regions (blue), and 29.2 million (−2.8%) in rural regions (green & purple).

Further changes in the transportation system such as infrastructure construction and public transit schedules are difficult to project, such that the transportation network reflects the current status of about 2020. Moreover, using a transportation network without changes allows description of the pathway for CAD more directly, as external factors on the network have no influence. The direct effects of CAD on the transportation system can be derived by comparing the reference to the automated scenarios described in the following sections.

(1) *Private car automation*

The scenario of private car automation *scenario* (1) extends the reference scenario by making vehicle automation available to private households. The models assume that in 2050 all car types could be upgraded with automation. Thus, in the model every household has the possibility of choosing between a private conventional vehicle (PCV) or a private automated vehicle (PAV). In other words, as the automation of the private car does not seem reversible, future scenarios should always include PAV. In terms of space, the models assume that PAV can travel everywhere, like PCV. So, in the case of private car automation, there are no regional and/or infrastructural limitations.

(2) *SAV in urban and rural regions*

The scenario of SAV introduces an additional shared vehicle service as a transportation mode. First, users request a trip from A to B. Then, a dispatcher sends a vehicle to serve the request. The period between the request and the vehicle's arrival is defined as waiting time. The user can book SAV as car-sharing (SAVc) when the user wants to ride alone, and in this mode the vehicle travels on the fastest route between the origin and the requested destination. Or users can share their SAV as ride-sharing (SAVr), such that multiple users with different start and end locations might share a vehicle if the detour distance and additional travel time are within an acceptable tolerance. Thus, SAVr increases the travel time, but as a benefit, the users can share the costs to reduce travel expenses.

Regarding market entry, it seems realistic to assume that SAV will begin its operation in limited areas. Nevertheless, it is unclear if these areas will lie in an urban context with complex traffic situations but high demand, leading to higher revenue. Alternatively, rural regions provide less-complex traffic situations but might need new regulations in order to operate as a supplement to public transit. To evaluate the different implementation pathways of SAV, scenario (2) splits into two subsets. The first SAV scenario, "SAV in urban regions" *scenario (2a)*, refers to all urban regions with more than 100,000 inhabitants. In these areas, the SAV serve requests door-to-door, fulfilling the users' specific requests. Thus, the service is in direct competition with public transit. In the second SAV scenario, "SAV in rural regions" *scenario (2b)*, the vehicles only act as a so-called feeder to the railway. In other words, users must start or end their SAV trips at railway stations. Therefore, it is questionable if such a service is viable concerning user acceptance and cost-efficiency. Nevertheless, this scenario allows insights into the regulation of automated services and the effect of supplement services on public transit.

(3) *Full automation*

The third scenario allows for full implementation of PAV and SAV *scenario (3)*. Thus, SAV operates door-to-door and as a feeder in all urban and rural regions in Germany. This scenario is presumably not very cost-efficient, but it provides an upper boundary of CAD affecting the transportation system.

3.2 Model Overview

The topic of CAD is complex as it affects mobility and transportation in various ways. Concerning the transportation system, the main drivers are the improvement of private car travel and the introduction of new transportation modes such as SAV. Both aspects fundamentally influence various long- and short-term decisions, making traveling easier. As models often either simulate long- or short-term decisions, missing information is concluded based on assumptions. By combining different models, it is possible to replace assumptions with more accurate simulation results. Such a modeling framework can simulate the various changes due to CAD.

The proposed framework consists of three models (see Fig. 2):

- a car ownership model (COM) estimating car densities

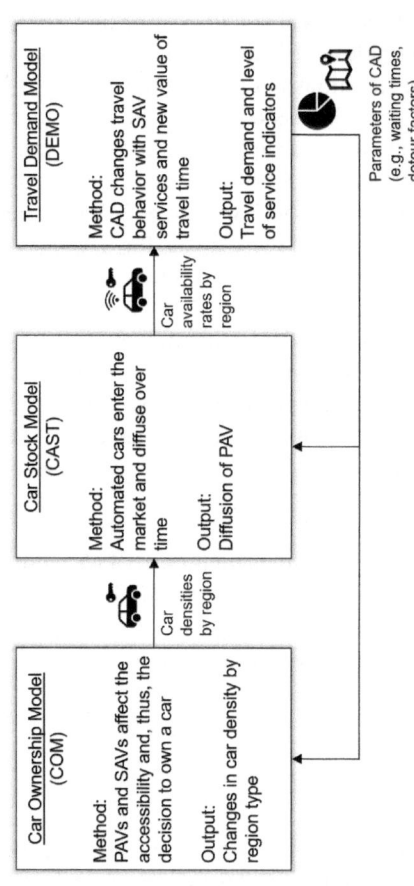

Fig. 2 Overview of model interaction

- a car stock model (CAST) adding diffusion of automation to the car densities
- a travel demand model (DEMO) using the level of automation of the fleet to simulate the interaction within the transportation system.

First, the car ownership model reflects a household's decision on whether to own a car or not. Besides socioeconomic attributes and the built environment, it includes accessibility attributes for various transportation modes. The implementation of PAV and SAV affects these accessibilities, resulting in changed car densities for every region type. Second, the new car densities enter the car stock model. As cars with different levels of automation enter the market over time, the households decide on whether to buy an automated car or not. Thus, automated cars diffuse in the vehicle stock over time, estimating the number of PCV and PAV for every region type and year. The car ownership model (COM) and the car stock model (CAST) simulate long-term household decisions affected by automation. Third, the travel demand model (DEMO) uses the vehicle stock of the reference year 2050 as a basis to model short-term decisions in travel behavior. It further simulates changes in travel behavior due to CAD. The model output shows various effects on the transportation system, such as changes in mode choice and travel distances. As the travel demand model simulates the SAV, it quantifies parameters such as the waiting time for SAV or detours because of ride-sharing. These parameters return to the previous models as they influence, for example, the accessibility and decision to buy a car.

3.2.1 Car Ownership Model (COM)

The car ownership model (COM) estimates car availability at the household level for Germany [49]. It is the first model within the proposed framework as it reflects on the long-term household decision of how many cars to own. This information is fundamental when forecasting car types or travel demand as it provides the limits within which car type choices or mobility choices take place.

The COM predicts car availability as the ratio of cars to the number of household members owning a driver's license. It has three levels: "no car available"—for households without cars or without household members owning driver's licenses; "shared car"—for households where the number of cars is less than the number of driver's licenses, and; "exclusive car"—for households where the number of cars is equal to or greater than the number of driver's licenses. To forecast car availability for the households, the model uses two successive binary logit models. The first model differentiates between households with and without a car. The second model separates the households with cars into the levels of "shared" and "exclusive" car availability. Furthermore, car availability can be transformed into car density by aggregating the households by region type.

The COM includes demographic, economic, built environment, and transportation supply attributes, as these categories influence car ownership and car availability (see Fig. 3 and Schrömbges and Kuhnimhof [49]). The basis for the model is the German NHTS "Mobility in Germany 2017" [27], in which households report on their socioeconomic circumstances and their mobility attributes, such as car owner-ship. After filtering for complete reporting households, the final sample consists of 264,270 persons living in 122,467 households. The NHTS survey includes individual demographic information (e.g., age or gender), household demographic attributes (e.g., household size or presence of children), and economic attributes (e.g., economic status or the number of workers). Further attributes append to the dataset based on the geo-location of the households' residences. The built environment covers vari-ables such as population density or building density [13]. These serve as a proxy to describe the immediate surroundings of the households' locations. For example, high population and building densities indicate urban regions, whereas low values reflect rural regions. In addition to the immediate households' neighborhoods, acces-sibility to activity locations further away also affects car ownership. In the context of car availability in COM, accessibility means the number of inhabitants reach-able by the available transportation supply. This method is known as the cumulative opportunities approach, as it sums up all accessible inhabitants who live within a set travel time budget of the start location [4]. In order to limit calculation time, a budget of 30 min is set for all transportation modes in the reference scenario to compare between walk & bike (non-motorized individual transportation—NMIT), car (motorized individual transportation—MIT), and public transit (PT).

The transportation supply is the key determinant for the CAD scenarios, as it is the only category that reacts to changes in the transportation system due to PAV and SAV. For example, the automation of private cars leads to a reduction in the value of travel time as driving becomes more effortless and in-vehicle time can be spent otherwise. In the COM context, this leads to an increased time budget for accessibility by car. Nevertheless, one could assume presumably realistic values, or conduct dedicated choice experiments in order to quantify this effect. Other research studies expect the impact of automation on travel time savings to lie between 25 to 50% [35]. Kolarova et al. [36] notice an effect of 40% for commuters, but measure no significant change

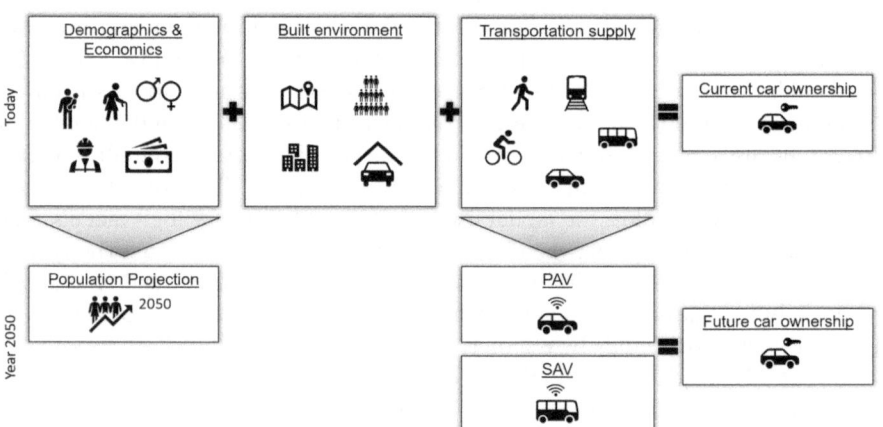

Fig. 3 Schematic methodology of modeling car ownership in the context of automation

for shopping or leisure trips. Therefore, the accessibility calculation of PAV uses a span of 20–25% for an increased travel time budget.

SAV, in the form of a door-to-door service, is expected to be a viable alternative to a private car as it offers high flexibility. SAV, in the form of a feeder to railway stations, is a supplement to public transit. Therefore, in the calculations the SAV accounts for public transit. SAV increases accessibility as it opens up new areas through direct connections. In other words, it is possible to reach more inhabitants within the same travel time budget.

Automated driving enters the COM by increasing the travel time budget for PAV and generating more direct connections via SAV. As a result, COM forecasts car availability, which enters the CAST by providing an upper boundary for car density by region type.

3.2.2 Car Stock Model (CAST)

The car stock model (CAST) provides data on the future development of the German passenger car fleet [32, 37]. CAST is used to map and evaluate the effects of transportation policy measures and changes in framework conditions on the passenger car fleet. One component of this is the diffusion of new technologies, e.g., automated vehicles and alternative powertrains. CAST output can be applied as an input for car availability by level of automation in different travel demand models.

CAST is a dynamic agent-based model that simulates the development of the German passenger car fleet in annual steps. The core of CAST is a decision model for car purchases by private households. The households are initially identified on the basis of the German NHTS "Mobility in Germany 2017" [27] and their composition is projected according to population forecasts of the Federal Statistical Office [11].

The decision model is composed of three discrete choice models that build on one other. Firstly, making use of a mixed logit model, households that buy a new or used vehicle in the respective year are identified. Secondly, used car buyers choose the quality of the vehicle according to preference of age and mileage of the vehicle, applying an ordered logit model. Thirdly, the car buyers choose the technology of the vehicle, which is a combination of the vehicle's size, powertrain, and automation level (see Fig. 4).

As part of the model's vehicle technology decision, households decide on the vehicle's automation level. This decision is affected by three factors. First, additional costs of PAV compared to PCV affect investment costs and, thus, the utility of opting for an automated vehicle. Second, the year of market entry restricts availability of Level 4+ automation to the years after 2035. Before 2035, Level 4+ is not included in the choice set of households. These two factors are model inputs provided by industry experts. Third, due to the lack of empirical data, the vehicle technology decision is also based on model assumptions that refer to prevailing reservations of households about the automation technology. These reservations may relate to safety or data protection concerns, for example, or the enjoyment of independent driving. It is assumed that reservations decline steadily over time.

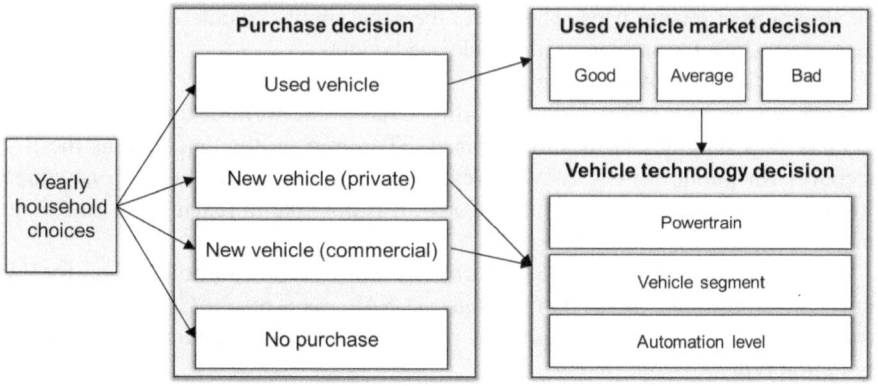

Fig. 4 Discrete choice modeling of passenger car purchases by households with CAST

CAST differs from other passenger car fleet models by integrating both the new and used car market. Since CAST does not model production volumes of vehicle manufacturers, the supply of PAV is not restricted on the new vehicle market. However, supply on the used vehicle market is restricted by the number of vehicles available each year. This number represents the sum of vehicles that enter the used vehicle market plus the unassigned vehicles every year (car dealer). If demand for a vehicle technology exceeds supply then the model applies an algorithm that alters the price of the specific vehicle technology incrementally and, in this way, changes the utility of this specific option. Consequently, some households will change their choice to another technology due to the higher price. This so-called shadow price methodology is used to achieve a fair distribution of vehicle technologies. This methodology is used twice, first for the size and powertrain combinations, and then for the automation levels, so that there are different shadow prices for each specification.

CAST models the vehicle purchases of private households. For the interface with the travel demand model DEMO, household information is disaggregated into person information referring to vehicle availability for every person in each household. Vehicle availability is modeled in such a way that all household members have access to a PCV or PAV, if such vehicles exist in the household. The CAST output (and DEMO input) is a dataset that contains information on how many persons in which region type of which age, gender, and employment have access to PCV and PAV per year until 2050.

3.2.3   Travel Demand Model (DEMO)

The German national travel demand model DEMO (DEutschland MOdell) is applied in order to forecast traffic flows and travel behavior depending on policy, socio-demographic, and technology developments [59]. There are modules for long-distance and short-distance passenger transportation. The effects of CAD are assessed for everyday trips, thus using the short-distance module. Travel demand in this module is determined using a four-step-approach, with trip generation, joint destination and mode choice, and a network assignment. The demand model was calibrated to represent mobility behavior observed in the German NHTS "Mobility in Germany 2017" [27]. It is expected that the introduction of CAD will influence travel behavior on different levels, as studies such as Kolarova et al. [36] suggest a shift in value of travel times with automation. Furthermore, SAV services provide new modes of transportation, which can influence accessibility and thus mode and destination choice preferences. Furthermore, diffusion rates of automated vehicles and general car ownership are important input values, which are influenced by the introduction of CAD. These changes are implemented in DEMO with an upstream car ownership model step, using COM and CAST.

In DEMO, car ownership is represented by a car availability rate for each transportation analysis zone, which is initially calibrated using the results of the NHTS "Mobility in Germany 2017". This variable describes the share of people with access to a private car in their household. With this method, regional differences in motorization can be represented in the model. The impact of CAD on car ownership determined in COM and CAST is included with an incremental adjustment of these availability rates, with different growth rates depending on regional types. These growth rates are derived by comparing car ownership in the different diffusion pathways to a baseline scenario without automation. This is however not the only adjustment regarding car availability, as the introduction of CAD also introduces a new car type with different choice parameters.

Since car ownership, and thus the decision for PAV, is a long-term household decision, this choice is not a logical part of a standard mode choice model. Therefore, the resulting diffusion of PAV among car owners is calculated in CAST and included as an input condition for a restructured demand model in DEMO, as shown in Fig. 5. These diffusion rates are used to distribute generated individual trips on two specialized mode choice models, where either PCV or PAV represent the car alternative. Mode and destination choice are calculated for both sub-models with the given total number of trips, as further described in Winkler and Mocanu [58]. Travel times and costs are important indicators for the formulation of utility. For automated driving, travel times are reduced using the findings of Steck et al. [52]. Empty trips from automated parking and relocation of vehicles are not regarded.

In order to include SAV into the demand model, new transportation modes are created. As stated in the description of scenario (2), the assumption is that a commercial service provider with a fleet of SAV offers taxi services with a ride-sharing option. In the demand model, this results in two options: either use SAV to travel alone (SAVc) or with other passengers (SAVr). The parameters and components for

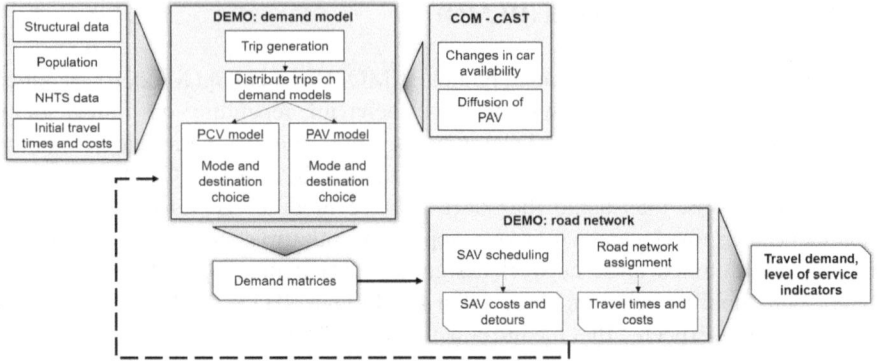

Fig. 5 Schematic overview of modeling CAD and SAV in the macroscopic travel demand model DEMO

the formulation of utility are determined as a mix between car and public transit. The main differences in the utility components are costs and detours, with SAVr trips having a lower cost due to their higher occupancy rates, and SAVc trips not having any detours. The costs and detours are calculated with a specialized SAV algorithm in the network model.

During the SAV algorithm, the first step is to match SAVr trips with similar directions using the approach of Friedrich et al. [17], where trip trajectories are compared by the zones they cross. This approach is adjusted for DEMO, as the granularity does not allow for a detailed depiction of trajectories with the base approach. Instead of the zones themselves, population centroids for each zone are used as pick-up locations, further allowing for the inclusion of inner-cell trips and calculation of detours [54]. As a result, vehicle trips for the calculated SAVr demand are determined. Afterwards, a scheduling algorithm is applied for the sum of SAVc and SAVr vehicle trips to determine fleet size and relocation trips [23, 46]. For cost calculation, further input by industry experts is used concerning fixed and running costs (see chapter "Business Analysis and Prognosis Regarding the Shared Autonomous Vehicle Market in Germany").

To assess the effects of CAD on transportation volumes and travel times, a network assignment is performed using all private car trips (conventional and automated), SAV trips and fixed road volumes for long-distance car trips and freight transportation. In an iterative process, mode and destination choice steps are repeated until there are no significant changes in travel times to the previous iteration. As a result, DEMO produces several indicators to evaluate the effects of CAD in road-based passenger transportation, including modal splits, vehicle mileage and travel times on a national and regional level.

4 Effects of Automation on the Transportation System

When applying the CAD scenarios in the model framework, diverse changes in the transportation system occur. Thereby, each model results in different outputs. Whereas COM primarily assesses changes in car density at household level and CAST concerns the fleet composition, DEMO is able to analyze the most effects as it is a full macroscopic model.

4.1 Effects of CAD on Car Ownership Due to Changes in Accessibility (COM)

The car ownership model simulates the emergence of CAD by adapting the accessibilities at the home location of the surveyed households. Therefore, a detailed analysis of the accessibility changes due to CAD is necessary to evaluate the model outcome of car densities. In total, the analyses include approximately 43,000 different households' home locations, with approximately 13,000 located in urban regions and 30,000 in rural regions (see Fig. 1).

For traveling by car, automation increases the accessibility by PAV due to the increase in travel time budget compared to PCV in the reference 2050 (see Methodology Sect. 3). For the scenario of private automation, that means all households have the accessibility potential provided by PAV, irrespective of their likelihood of adopting or owning a PAV. The theoretical background for this assumption is that households consider the additional accessibility benefits of PAV when purchasing a new vehicle. The travel accessibilities by public transit are linked to the accessibilities by SAV, leading to more direct connections (see Methodology Sect. 3). Again, the assumption is that SAV are fully accessible and available, resulting in a layout in which all households can afford to use SAV, and the fleet is large enough to fulfill all requests at any time. Nevertheless, the accessibilities include region-based waiting times and detour factors to make the calculations more realistic.

Figure 6 provides an overview of the relative changes in accessibility by urban and rural regions (columns) and scenarios of automation (rows). Each graph splits further into the four region types and a total value for all locations (see Fig. 1 for spatial regions). For every location, the improvement through automation leads to a higher number of accessible inhabitants. This improved accessibility is compared to the accessible inhabitants of the reference 2050. In other words, a value of one indicates no improvement, whereas a value of two refers to a doubling of accessible inhabitants.

The scenario (1) Private car automation (row 1) has only an improvement of 1.5 on average, whereas central rural regions profit most with 1.8 on average. The slight improvement in accessibility comes from the fact that the car already reaches a large number of residents within the assumed travel time budget of 30 min. Therefore, automation only contributes a little to the number of accessible inhabitants. The

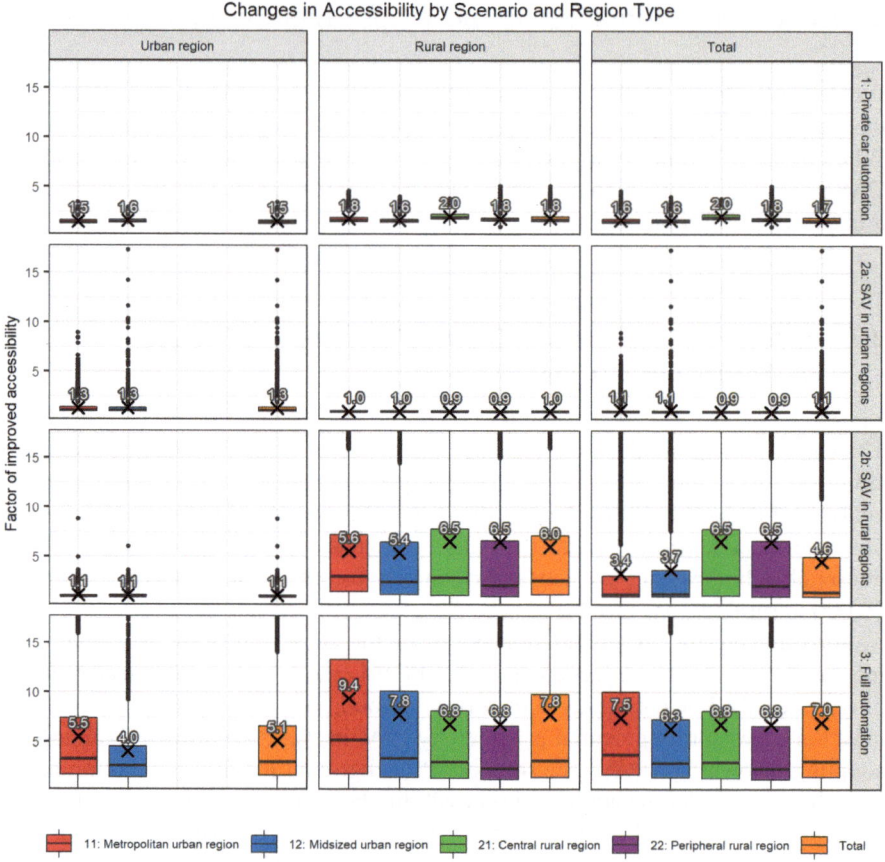

Fig. 6 Relative improvement of accessibility by region type and scenario (for readability upper boundary set at factor 17)

scenario (2a) SAV in urban regions (row 2) shows the improvement in accessibility due to the door-to-door service. As the left column shows the urban regions, it includes all locations where the service operates. However, it provides only a minor benefit as public transit is already of a good quality concerning accessible inhabitants in urban regions. The deviation from one for rural regions is due to errors in locating households limited by data privacy. The scenario (2b) SAV in rural regions (row 3) introduces a SAV as a feeder service to railway stations in rural regions. The direct access to the railway leads to a factor of about six times more accessible inhabitants via public transit. Urban regions inherit a slight improvement as traveling from urban to rural regions allows using the feeder service for egress from the railway stations. Scenario (3) Full automation (row 4) implements PAV and SAV as a door-to-door service and a feeder service in all regions. This scenario shows the upper boundary in accessibility gained through automation. Interestingly, urban regions benefit by a factor of approximately five less than rural regions, which differ between seven to ten. In detail, rural regions located close to metropolitan or midsized regions benefit more than central or peripheral rural regions as the population density decreases, such that fewer inhabitants are accessible within the given travel time budget.

These changes in accessibility enter the COM resulting in new probabilities of car availability and car densities. By comparing the densities of the reference scenario to the new densities, the impact of CAD is visible (see Fig. 7). The automation of the private car leads to an average increase of about 1% in car ownership. Providing a door-to-door service in urban regions results in a reduction of −3.6% on average. Adding a feeder system in rural regions reduces car ownership by −2.8%. The combination of PAV with full SAV reduces car densities on average by approx. − 3.9%.

Despite the considerable improvement in accessibility, the effect on car densities due to CAD is only slight. By interpreting the COM model in detail, the conclusion is that socioeconomic factors such as changes in life circumstances or availability of income dominate the decision to own a car. Accessibility only contributes to minor changes in car densities. Nevertheless, small changes in car ownership due to CAD might affect the outcome of the car stock model (CAST) and travel demand model (DEMO).

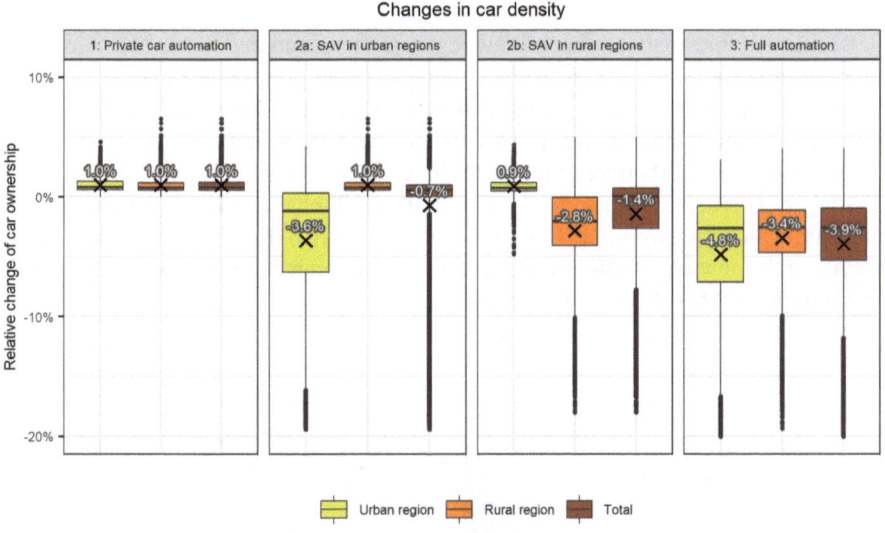

Fig. 7 Relative changes in car densities

4.2   Nationwide Assessment of Diffusion of Automated Vehicles (CAST)

In reference to the model descriptions above, the results of the effect of CAD on car densities in the respective scenarios by the COM enter into the model CAST that projects the share of PAV in the passenger car fleet. Fleet modeling results by CAST are based on three major inputs:

- the year of market entry of PAV and the additional costs of PAV compared to PCV vehicles, which comes from industry experts,
- car densities for the year 2050 for every scenario, which is the COM output,
- model assumptions on dynamic parameter changes that affect the utility of PAV. Those parameters pool all additional factors that hamper the purchase of PAV by households or companies. Such factors may relate to safety or data protection concerns, for example, or the enjoyment of independent driving.

Based on these inputs, CAST shows an uptake of PAV market shares (Level 4+) starting from 2035. Because of additional costs of and reservations about the technology, initial market shares of PAV are low. Mostly due to fading reservations caused by technology acceptance, market shares increase steadily and reach 70% in 2050.

Because of the small number of new registrations compared to total vehicle stock, the ramp-up of PAV in the German passenger car fleet progresses rather slowly. The duration that passenger cars stay in the fleet is modeled endogenously by CAST. On average, this modeled duration is 15 years for both PCV and PAV. As PAV market shares increase steadily over time, stock shares grow faster every year. However, the passenger car stock for the scenario year 2050 is still characterized by a mix of PCV and PAV, which then make up 44% of the stock. Figure 8 illustrates the yearly shares of PAV in the new registrations and the stock of passenger cars as modeled by CAST for scenario (1) Private car automation.

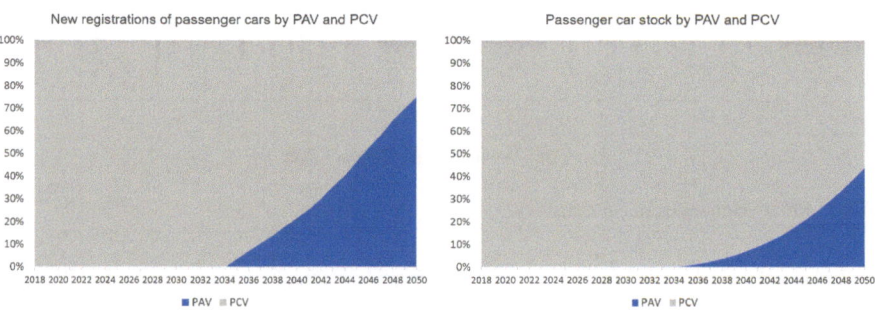

Fig. 8 New registrations and stock of passenger cars in Germany by PCV and PAV (CAST modeling, scenario (1) Private car automation)

While the year of market entry and the additional costs of PAV as well as the assumptions on fading reservations caused by technology acceptance are the same for all scenarios, the car densities modeled by COM differ slightly between the scenarios. Consequently, diffusion rates of PAV in the passenger car stock of 2050 differ only marginally between the scenarios. The diffusion rates for all scenarios are:

- 44.0% for scenario (1) private car automation,
- 43.5% for scenario (2a) SAV in urban regions,
- 43.3% for scenario (2b) SAV in rural regions,
- 42.8% for scenario (3) full automation.

The PAV and PCV in the 2050 fleet are transformed into car availability of individual persons disaggregated by region type, age, sex, and employment and, in this form, represent the input data for the travel demand model DEMO.

4.3   Effects of CAD on Travel Demand (DEMO)

The travel demand model DEMO was applied for three diffusion pathway scenarios:

- (0) Reference 2050
- (1) Private car automation
- (2a) SAV in urban regions.

In order to optimize computational effort while still modeling SAV, the scenarios SAV in rural regions (2b) and full automation (3) were omitted in DEMO. Additionally, scenario (2a) already covers large amounts of everyday travel, as trips within cities with SAV services make up 42% of all model trips, as shown in Fig. 9. The spatial distribution of model trips among the regional classification from Fig. 1 is also shown in Fig. 9.

Figure 10 shows key model results of DEMO for Germany, including modal splits, mileages and travel time changes. Looking at modal splits, there is a small increase in car trips with the introduction of PAV. Private conventional and automated cars are roughly used for the same number of trips. Adding SAV in urban regions however,

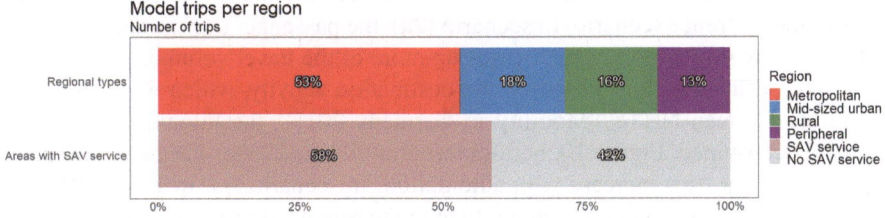

Fig. 9 Share of trips per region in DEMO

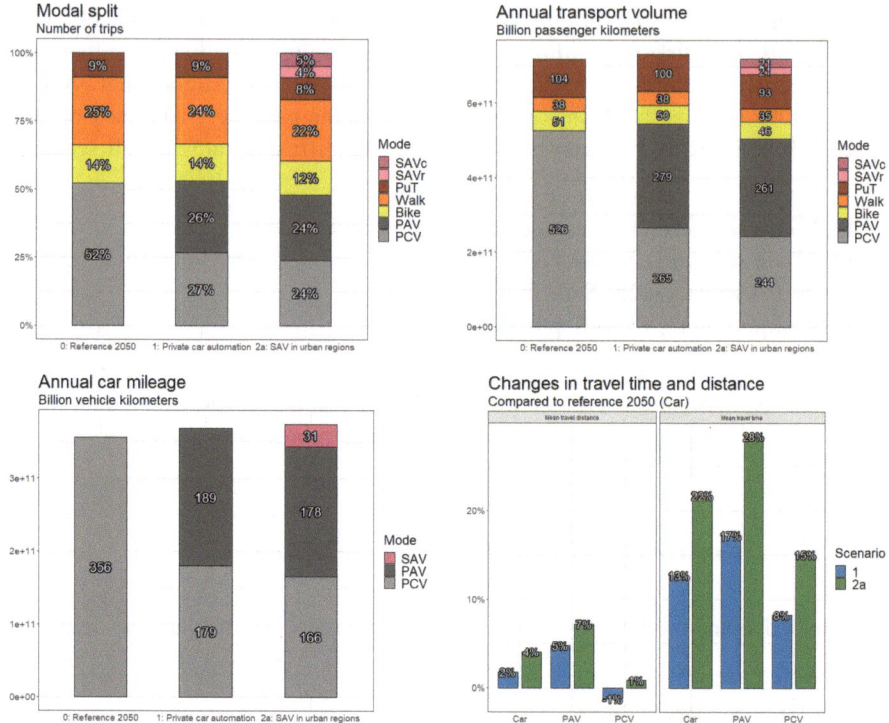

Fig. 10 DEMO results for Germany

leads to a decrease in car trips, as trips are replaced with new mobility options. In general, all modes suffer decreases in trips by 8–10%.

Comparing passenger transportation volume, the total volume faces a small increase with PAV and is similar to the reference with SAV in urban regions. The mean distances traveled by car are slightly higher in the automated scenarios, with the mean distance however staying similar over all trips. The combination of small increases in trip distance and number of trips for cars leads to a 3% increase in passenger kilometers traveled using cars in scenario (1). As the modeling approach using perceived travel times suggests, longer distances are more attractive with a PAV, thus PAV trips in the automated scenarios are around 5–7% longer than car trips in the reference scenario. In scenario (2a), the passenger transportation volume for cars decreases by 4%, as SAV takes up more of the travel volume.

Using occupancy rates based on the destination activity, which were calibrated using the German NHTS "Mobility in Germany 2017", the kilometers driven by cars are determined. Figure 10, bottom left plot, illustrates that the annual mileages of everyday car trips increase with automation. In scenario (1), with only PAV, the total car mileage is around 3% higher than in the reference scenario. The kilometers driven by PAV make up 51% of the total mileage. Similar to the passenger kilometers, vehicle mileage of PAV for the scenario with SAV in urban regions is roughly 4%

lower than the reference scenario. The total annual mileage is however highest for this scenario when SAV are included, with an increase of 5% to the reference and 2% to private automation only.

The mean travel time per car trip, as shown in the bottom right plot in Fig. 10, increases by roughly 13% in scenario (1), which can be contributed to the changes in perceived travel time with PAV and congestion effects. The increase for PCV trips only is lower with 8%. The introduction of SAV has an even stronger effect on travel times for cars, as there is a 21% increase for cars in scenario (2a) and a 15% increase for PCV only. In comparison, the mean distance per PCV trip is roughly the same in all three scenarios. This indicates that the additional vehicle volume due to SAV leads to higher congestion in the model.

Automation also has an effect on the road network, with changing traffic loads and subsequently, effects on travel times due to congestion effects. Figure 11, left plot, shows increases in mileage per square kilometer in the east and metro-regions around Berlin, Munich and Hamburg. For the remaining country, traffic volumes remain similar. Adding SAV, higher traffic loads in urban regions with these services can be observed due to SAV despite a lower modal share of car trips.

SAV fleet sizes, occupancy rates and cost per kilometer were determined for each urban region and service area with a scheduling algorithm for both SAVc and SAVr trips, as well as input from an industry analysis (see chapter "Business Analysis and Prognosis Regarding the Shared Autonomous Vehicle Market in Germany"). It is

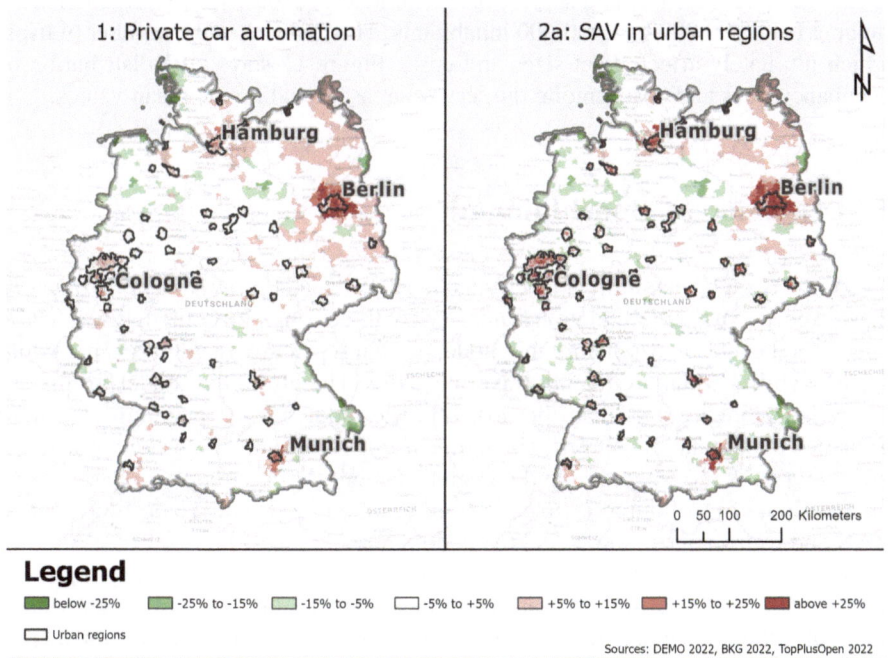

Fig. 11 Changes in mileage per square kilometer in the automated scenarios

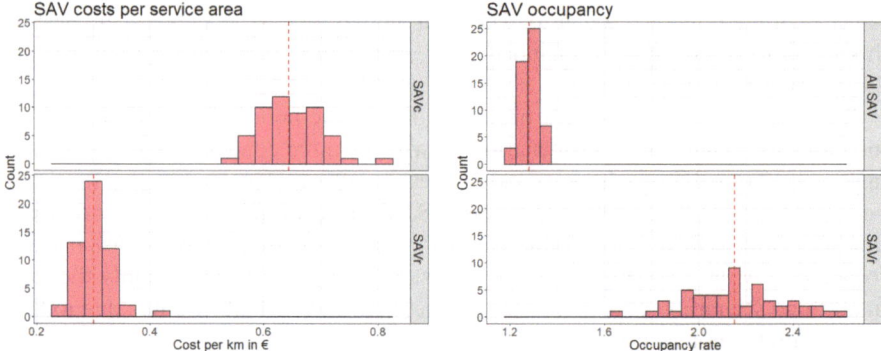

Fig. 12 Distribution of cost per km and occupancy rates for (2a) SAV in urban regions in the service areas

assumed that the fee for SAVr trips is reduced by dividing the cost per kilometer by the occupancy rate. Overall, occupancy rates for SAVr trips range from 1.65 to 2.59, while overall occupancy including SAVc trips is between 1.19 and 1.36 (see Fig. 12). The fleet sizes are between 7.9 and 26.4 vehicles per 1,000 inhabitants, including a 10% reserve. This leads to costs of 0.55 €/km in the Rhine-Ruhr metropolitan region, including cities like Dortmund and Düsseldorf, up to 0.80 €/km in the mid-sized city Würzburg. The differences in SAV indicators between service areas can be attributed to the fact that the cities and service areas are heterogenous, with population sizes ranging from 100,000 to 4,000,000 inhabitants. This influences the number of trips, which ultimately affects fleet sizes and costs. Figure 12 shows the distribution of occupancy rates and costs among the service areas, including the mean values.

5 Discussion of Model Limitations

CAD affects car ownership, car stock, and travel demand in various ways. To quantify these aspects, a unique simulation framework of three consecutive models was developed. The benefit of combining the models lies in replacing assumptions regarding car availability when simulating travel behavior. Therefore, the model results are somewhat more realistic than those of separate models when planning for future scenarios of automation. Nevertheless, forecasting future situations contains several uncertainties as the real world and human behavior are very complex. These uncertainties come from limited or uncertain information on the future development of CAD and restrictions in the model methodologies.

Although the model framework reduces the number of assumptions by replacing them with modeled estimates, various influencing variables are still unknown. In terms of technology, it is uncertain when the different levels of automation will enter the market. Research studies vary between estimates of around 2025 and 2050, or even later for the implementation of PAV [55]. Within the car stock model (CAST), the market entry is set to 2035, leading to a vehicle stock of 44% automated vehicles in 2050. The earlier the market entry happens, the higher the share of PAV, the modal share, and vehicle kilometers—and vice versa.

Economic prognoses vary, referring to the additional costs of PAV and realistic costs of SAV for fleet operators. Concerning the latter, the chapter "Business Analysis and Prognosis Regarding the Shared Autonomous Vehicle Market in Germany" provides industry analyses and prognoses for the German market. For the purchase or leasing of private- or company cars, the cost of automation technology will make up considerable shares of vehicle prices, especially in the small and medium segments. As the willingness to pay greatly influences the adoption of automated vehicles [21], the model results react sensitively to different cost scenarios. Fewer costs will increase the speed of the market entry of PAV, such that the described results are likely to occur earlier than 2050.

Costs are not the only factor when speaking about the acceptance of PAV and SAV (e.g., see Literature review or chapter "Social Acceptance of CAD in Japan and Germany: Conceptual Issues and Empirical Insights"). It is impossible to accurately predict the uptake due to the complex interactions between the various acceptance determinants. To simplify this circumstance, the model framework assumes that all persons and households consider buying or riding automated vehicles. As a matter of fact, this assumption results in an overestimation of the effects of CAD as market shares and mode usage are presumably lower. Therefore, the proposed results should be considered as an upper boundary for the effects of CAD on transportation in Germany.

Furthermore, the simulation framework includes several simplifications to keep the models' interactions and calculation times within acceptable complexity. Accessibilities changing due to CAD are the key component of the car ownership model (COM). There are various methodologies to measure accessibility [4], but multiple analyses showed no considerable sensitivity in the model outcome. Thus, the current model version only uses cumulative opportunities, neglecting aspects such as accessible jobs or travel costs. Changes in accessibility might also result in household relocation. For example, if traveling by PAV will be easier than by PCV, people will live further away from the cities, resulting in urban sprawl and more vehicle kilometers in the transportation system [43]. However, this complex interaction does not enter the model framework due to simplification.

In the model framework, various choices concerning car ownership, car type, and mode usage are simulated. CAD adds new components to the choice sets. Observable and quantifiable variables such as travel times are possible to project, and thus form part of the model. However, unobservable attributes like safety concerns, or uncertain variables like safety distances between vehicles, which might either decrease through communication between the vehicles or increase due to safety concerns [24, 50], are neglected in the simulations. Therefore, the results inherit an uncertainty.

The model framework requires an interchange of information between the three models. As each model considers different input data concerning space and time, it is necessary to aggregate information. This step allows the combination of the models but also reduces the accuracy of the results. Nevertheless, the proposed model framework provides accurate results, as the key component of car availability is not assumed but modeled with respect to changes due to CAD.

6 Conclusion

Connected and automated driving (CAD), in the form of PAV and SAV, will transform the transportation system. Despite extensive research and technical testing, the impact of CAD is still unclear. The advantages, like a presumable reduction of car ownership or increased accessibility, oppose assumptions of increased urban sprawl and congestion. Thereby, the size of influence in each direction largely depends on the future concept of CAD. Economic considerations, societal implications, environmental restrictions, and political regulations build the framework for CAD. Thus, it is necessary to provide detailed knowledge about the impact of CAD in order to foster beneficial decisions.

The current study contributes to forecasting the impact of CAD on the transportation system by providing a simulation framework for Germany. It contains three consecutive models of car ownership (COM), car stock (CAST), and travel demand (DEMO). This model set allows combining long-term decisions of car ownership and type of automation with short-term decisions of travel behavior such as mode choice. Therefore, the simulation framework considers multiple effects of automation.

A total of four scenarios including automation are simulated to evaluate the results of different pathways. The first scenario analyses the influence of private car automation. It indicates only a minor increase of about 1% in car ownership. This small growth can be explained as accessibility by car is already very good, such that automation provides only a small additional benefit. Thereby, rural regions profit more than urban regions. Overall, the 2050 vehicle stock consists of 44% PAV. Thus, PCV and PAV drive as mixed traffic in the streets. Therefore, PAV cannot realize their full potential regarding safety and efficiency. In terms of the transportation system, PAV travel over 50% of the kilometers driven, which translates to an increase of 3% in total mileage. Overall, private car automation has a minor impact on car stock and ownership, but still leads to a disproportionate increase in vehicle kilometers driven. Further regulation could focus on discouraging driving longer distances.

The second scenario simulates automated door-to-door services in urban regions. The accessibility in these regions improves only slightly as public transit is already of good quality. Nevertheless, the number of private vehicles in urban regions reduces by 3.6% on average. However, the new transportation mode compensates for this effect by introducing new vehicles. As this service operates additionally in the transportation system, it attracts trips from all other existing means of transportation. As a result, the vehicle kilometers traveled increase by about 5%, which, especially in cities, might lead to further congestion. Therefore, introducing SAV should be closely monitored, and, if necessary, regulations such as limits on fleet size or operating hours should be implemented. A variation of the second scenario evaluates the

SAV fleet as a feeder to railway stations in rural regions. Such a service improves accessibility, especially in rural regions, but nearby urban regions also profit from better accessibility to their surroundings. Consequently, car ownership decreases in these regions by almost 3% on average. However, further analyses are necessary to evaluate whether the costs for such shared services outweigh the small benefits concerning reduced car ownership.

The third scenario implements full automation of PAV and SAV all over Germany. Therefore, it serves as the upper boundary of impacts due to CAD. Although vehicle ownership decreases on average by approximately 4%, the additional vehicles for the shared fleet might negate the reduction of private vehicles. Further research should analyze the cost structure in more detail.

The final model results forecast only a minor effect of CAD in reducing car ownership, and at the same time, an increase in SAV results in additional vehicle mileage in the transportation system. In order to create advantages from CAD, such as guaranteeing accessibility to further user groups including children, or persons with reduced mobility, and at the same time providing efficient transportation systems without increasing travel times due to congestion, further regulations on fleet size, operating hours, and areas are necessary. This research continues by modeling other regulatory frameworks, especially under the context of climate change.

References

1. Acheampong, R. A., & Cugurullo, F. (2019). Capturing the behavioural determinants behind the adoption of autonomous vehicles: Conceptual frameworks and measurement models to predict public transport, sharing and ownership trends of self-driving cars. *Transportation Research Part F: Traffic Psychology and Behaviour*, 62(4), 349–375. https://doi.org/10.1016/j.trf.2019.01.009.
2. Anowar, S., Eluru, N., & Miranda-Moreno, L. F. (2014). Alternative Modeling Approaches Used for Examining Automobile Ownership: A Comprehensive Review. *Transport Reviews*, 34(4), 441–473. https://doi.org/10.1080/01441647.2014.915440
3. Bates, J., & Mick, R. (1981). The factors affecting household car ownership. Farnborough/Hants.: Gower.
4. BBSR (April 2019). Methodische Weiterentwicklungen der Erreichbarkeitsanalysen des BBSR (BBSR Online-Publikation No. 09). Bonn. Retrieved from https://www.bbsr.bund.de/BBSR/DE/veroeffentlichungen/bbsr-online/2019/bbsr-online-09-2019-dl.pdf?__blob=publicationFile&v=1.
5. BBSR (2021). Raumordnungsprognose 2040. Retrieved from https://www.bbsr.bund.de/BBSR/DE/veroeffentlichungen/analysen-kompakt/2021/ak-04-2021.html.
6. Beiker, S. A. (2015). Einführungsszenarien für höhergradig automatisierte Straßenfahrzeuge. In M. Maurer, C. J. Gerdes, B. Lenz, & H. Winner (Eds.), *Autonomes Fahren: Technische, rechtliche und gesellschaftliche Aspekte* (pp. 197–217). Berlin/Heidelberg: Springer.
7. Bischoff, J., & Maciejewski, M. (2016). Simulation of City-wide Replacement of Private Cars with Autonomous Taxis in Berlin. *Procedia Computer Science*, 83, 237–244. https://doi.org/10.1016/j.procs.2016.04.121.
8. BKG (2022). TopPlusOpen. Retrieved from https://gdz.bkg.bund.de/index.php/default/wms-topplusopen-wms-topplus-open.html.
9. BMDV (2021). Regionalstatistische Raumtypologie (RegioStaR). Retrieved from https://www.bmvi.de/SharedDocs/DE/Artikel/G/regionalstatistische-raumtypologie.html.

10. Daziano, R. A., Sarrias, M., & Leard, B. (2017). Are consumers willing to pay to let cars drive for them? Analyzing response to autonomous vehicles. *Transportation Research Part C: Emerging Technologies, 78,* 150–164. https://doi.org/10.1016/j.trc.2017.03.003.
11. Destatis (2022). 14. koordinierte Bevölkerungsvorausberechnung. Retrieved from https://www.destatis.de/DE/Themen/Gesellschaft-Umwelt/Bevoelkerung/Bevoelkerungsvorausbe rechnung/_inhalt.html.
12. Eurostat (2021). EUROPOP2019: Population projection. Retrieved from https://ec.europa.eu/eurostat/databrowser/view/tps00002/default/table.
13. Ewing, R., & Cervero, R. (2010). Travel and the Built Environment. *Journal of the American Planning Association, 76*(3), 265–294. https://doi.org/10.1080/01944361003766766.
14. Fagnant, D., & Kockelman, K. M. (2014). The Travel and Envinromental Implications of Shared Autonomous Vehicles, using Agent-based Model Scenarios. *Transportation Research Part C: Emerging Technologies, 40,* 1–13. https://doi.org/10.1016/j.trc.2013.12.001.
15. Fagnant, D. J., & Kockelman, K. M. (2018). Dynamic ride-sharing and fleet sizing for a system of shared autonomous vehicles in Austin, Texas. *Transportation, 45*(1), 143–158. https://doi.org/10.1007/s11116-016-9729-z.
16. Fraedrich, E., & Lenz, B. (2015). Vom (Mit-)Fahren: autonomes Fahren und Autonutzung. In M. Maurer, C. J. Gerdes, B. Lenz, & H. Winner (Eds.), *Autonomes Fahren: Technische, rechtliche und gesellschaftliche Aspekte* (pp. 687–708). Berlin/Heidelberg: Springer.
17. Friedrich, M., Hartl, M., & Magg, C. (2018). A modeling approach for matching ridesharing trips within macroscopic travel demand models. *Transportation, 45*(6), 1639–1653. https://doi.org/10.1007/S11116-018-9957-5.
18. Friedrich, M., Sonnleitner, J., & Richter, E. (2019). Integrating automated vehicles into macroscopic travel demand models. *Transportation Research Procedia, 41,* 360–375. https://doi.org/10.1016/j.trpro.2019.09.060.
19. Galich, A., & Stark, K. (2021). How will the introduction of automated vehicles impact private car ownership? *Case Studies on Transport Policy.* Advance online publication. https://doi.org/10.1016/j.cstp.2021.02.012.
20. Gerpott, T. J. (2021). Connected Car. In T. Kollmann (Ed.), *Handbuch Digitale Wirtschaft* (pp. 1071–1089). Wiesbaden: Springer Gabler.
21. Gkartzonikas, C., & Gkritza, K. (2019). What have we learned? A review of stated preference and choice studies on autonomous vehicles. *Transportation Research Part C: Emerging Technologies, 98,* 323–337. https://doi.org/10.1016/j.trc.2018.12.003.
22. Gruel, W., & Stanford, J. M. (2016). Assessing the Long-term Effects of Autonomous Vehicles: A Speculative Approach. *Transportation Research Procedia, 13,* 18–29. https://doi.org/10.1016/j.trpro.2016.05.003.
23. Hartleb, J., Friedrich, M., & Richter, E. (2022). Vehicle scheduling for on-demand vehicle fleets in macroscopic travel demand models. *Transportation, 49*(4), 1133–1155. https://doi.org/10.1007/s11116-021-10205-4.
24. Hartmann, M., Motamedidehkordi, N., Krause, S., Hoffmann, S., Vortisch, P., & Busch, F. (2017). Impact of Automated Vehicles on Capacity of the German Freeway Network. *ITS World Congress 2017 Compendium of Papes.*
25. Heinrichs, D. (2015). Autonomes Fahren und Stadtstruktur. In M. Maurer, C. J. Gerdes, B. Lenz, & H. Winner (Eds.), *Autonomes Fahren: Technische, rechtliche und gesellschaftliche Aspekte* (pp. 219–239). Berlin/Heidelberg: Springer.
26. Hossain, M. S., & Fatmi, M. R. (2022). Modelling the adoption of autonomous vehicle: How historical experience inform the future preference. *Travel Behaviour and Society, 26*(1), 57–66. https://doi.org/10.1016/j.tbs.2021.09.003.
27. Infas, DLR, IVT, & infas 360 (2019). Mobility in Germany: Short report. Bonn, Berlin.
28. ISO; SAE International. *Taxonomy and definitions for terms related to driving automation systems for on-road motor vehicles.* (Publicly Available Specification, ISO/SAE PAS 22736:2021(E)). ISO; SAE International: ISO; SAE International.
29. ITF (2015). Urban Mobility System Upgrade: How shared self-driving cars could change city traffic. https://doi.org/10.1787/5jlwvzdk29g5-en.

30. Jong, G. d., Fox, J., Daly, A., Pieters, M., & Smit, R. (2004). Comparison of car ownership models. *Transport Reviews, 24*(4), 379-408. https://doi.org/10.1080/0144164032000138733.
31. Jong, G. d., & Kitamura, R. (2009). A review of household dynamic vehicle ownership models: Holdings models versus transactions models. *Transportation, 36*, 743–773. https://doi.org/10.1007/s11116-009-9243-7.
32. Kickhöfer, B., Bahamonde-Birke, F. J., & Nordenholz, F. (2019). Dynamic modeling of vehicle purchases and vehicle type choices from national household travel survey data. *Transportation Research Procedia, 41*, 2–5. https://doi.org/10.1016/j.trpro.2019.09.002.
33. Kickhöfer, B., & Brokate, J. (2017). Die Entwicklung des deutschen Pkw-Bestandes: Ein Vergleich bestehender Modelle und die Vorstellung eines evolutionären Simulationsansatzes. *Zeitschrift für Verkehrswissenschaft.*
34. Kim, S. H., Mokhtarian, P. L., & Circella, G. (2020). Will autonomous vehicles change residential location and vehicle ownership? Glimpses from Georgia. *Transportation Research Part D: Transport and Environment, 82*, 1–17. https://doi.org/10.1016/j.trd.2020.102291.
35. Kolarova, V. (2021). Measuring, analysing and explaining the value of travel time savings for autonomous driving.https://doi.org/10.18452/23077.
36. Kolarova, V., Steck, F., & Bahamonde-Birke, F. J. (2019). Assessing the effect of autonomous driving on value of travel time savings: A comparison between current and future preferences. *Transportation Research Part A: Policy and Practice, 129*, 155–169. https://doi.org/10.1016/j.tra.2019.08.011.
37. Kröger, L., Kickhöfer, B., Bahamonde-Birke, F. J., Nordenholz, F., & Bolz, M.-S. (2018). Dynamic simulation of the German vehicle market. *7th Symposium of the European Association for Research in Transportation - hEART 2018*, Athens, Greece.
38. Kröger, L., Kuhnimhof, T., & Trommer, S. (2016). Modelling the Impact of Automated Driving - Private Autonomous Vehicle Scenarios for Germany and the US. *European Transport Conference 2016*, Barcelona, Spain.
39. Kröger, L., Winkler, C., & Kuhnimhof, T. (2018). Stepwise modelling of the impact of autonomous vehicles on the transport system in Germany for different scenarios. *European Transport Conference ETC 2018*, Dublin, Irland.
40. Lavieri, P. S., Garikapati, V. M., Bhat, C. R., Pendyala, R. M., Astroza, S., & Dias, F. F. (2017). Modeling Individual Preferences for Ownership and Sharing of Autonomous Vehicle Technologies. *Transportation Research Record: Journal of the Transportation Research Board, 2665*(1), 1–10. https://doi.org/10.3141/2665-01.
41. Lenz, B., & Fraedrich, E. (2015). Neue Mobilitätskonzepte und autonomes Fahren: Potenziale der Veränderung. In M. Maurer, C. J. Gerdes, B. Lenz, & H. Winner (Eds.), *Autonomes Fahren: Technische, rechtliche und gesellschaftliche Aspekte* (pp. 175–195). Berlin/Heidelberg: Springer.
42. Litman, T. (2022). Autonomous Vehicle Implementation Predictions. *Victoria Transport Policy Institute*. Retrieved from https://www.vtpi.org/avip.pdf.
43. Llorca, C., Moreno, A., Ammar, G., & Moeckel, R. (2022). Impact of autonomous vehicles on household relocation: An agent-based simulation. *Cities, 126*(2), 1–15. https://doi.org/10.1016/j.cities.2022.103692.
44. Maurer, M. (2015). Einleitung. In M. Maurer, C. J. Gerdes, B. Lenz, & H. Winner (Eds.), *Autonomes Fahren: Technische, rechtliche und gesellschaftliche Aspekte* (pp. 1–8). Berlin/Heidelberg: Springer.
45. Pertschy, F. (2022). Autonomes Fahren auf Level 5 ist nicht möglich. *AutomotiveIT*. Retrieved from https://www.automotiveit.eu/technology/autonomes-fahren/autonomes-fahren-auf-level-5-ist-nicht-moeglich-825.html.
46. Richter, E., Friedrich, M., Migl, A., & Hartleb, J. (2019). Integrating ridesharing services with automated vehicles into macroscopic travel demand models. In *Mt-ITS 2019: 6th International Conference on Models and Technologies for Intelligent Transportation Systems: Cracow University of Technology, 5–7 June 2019, Kraków, Poland* (pp. 1–8). Piscataway, NJ: IEEE. https://doi.org/10.1109/MTITS.2019.8883315.

47. Schoettle, B., & Sivak, M. (2015). Potential Impact of Self-Driving Vehicles on Household Vehicle Demand and Usage.
48. Schreurs, M. A., & Steuwer, S. D. (2015). Autonomous Driving-Political, Legal, Social, and Sustainability Dimensions. In M. Maurer, C. J. Gerdes, B. Lenz, & H. Winner (Eds.), *Autonomes Fahren: Technische, rechtliche und gesellschaftliche Aspekte* (pp. 151–173). Berlin/Heidelberg: Springer.
49. Schrömbges, M., & Kuhnimhof, T. (2023). Modeling the impact of accessibility on car availability in a nationwide model of Germany. In *102nd Annual Meeting of the Transportation Research Board, 8-12 January 2023, Waschington, D.C., USA.*
50. Sonnleitner, J., Friedrich, M., & Richter, E. (2021). Impacts of highly automated vehicles on travel demand: Macroscopic modeling methods and some results. *Transportation*, 1–24. https://doi.org/10.1007/s11116-021-10199-z.
51. Soteropoulos, A., Berger, M., & Ciari, F. (2019). Impacts of automated vehicles on travel behaviour and land use: an international review of modelling studies. *Transport Reviews*, *39*(1), 29–49. https://doi.org/10.1080/01441647.2018.1523253.
52. Steck, F., Kolarova, V., Bahamonde-Birke, F., Trommer, S., & Lenz, B. (2018). How Autonomous Driving May Affect the Value of Travel Time Savings for Commuting. *Transportation Research Record: Journal of the Transportation Research Board*, *2672*(46), 11–20. https://doi.org/10.1177/0361198118757980.
53. Stinson, M., Zou, B., Briones, D., Manjarrez, A., & Mohammadian, A. (2021). Vehicle ownership models for a sharing economy with autonomous vehicle considerations. *Transportation Letters*, *1285*, 1–17. https://doi.org/10.1080/19427867.2021.2007681.
54. Thomsen, N. (2023). Implementing a Ride-sharing Algorithm in the German National Transport Model (DEMO). *Transportation Research Record: Journal of the Transportation Research Board*, *2677*(5), 1–10. https://doi.org/10.1177/03611981221127015.
55. Ulrich, C., Frieske, B., Schmid, S. A., & Friedrich, H. E. (2022). Monitoring and Forecasting of Key Functions and Technologies for Automated Driving. *Forecasting*, *4*(2), 477–502. https://doi.org/10.3390/forecast4020027.
56. VDV (2020a). Autnome Shuttle-Bus-Projekte in Deutschland. Retrieved from https://www.vdv.de/liste-autonome-shuttle-bus-projekte.aspx.
57. VDV (2020b). New Mobility-Projekte in Deutschland. Retrieved from https://www.vdv.de/new-mobility-projekte.aspx.
58. Winkler, C., & Mocanu, T. (2017, October 4). Methodology and Application of a German National Passenger Transport Model for Future Transport Scenarios. *Proceedings of the 45th European Transport Conference*, Barcelona.
59. Winkler, C., & Mocanu, T. (2020). Impact of political measures on passenger and freight transport demand in Germany. *Transportation Research Part D: Transport and Environment*, *87*, 102476. https://doi.org/10.1016/j.trd.2020.102476.
60. Zhang, W., Guhathakurta, S., & Khalil, E. B. (2018). The impact of private autonomous vehicles on vehicle ownership and unoccupied VMT generation. *Transportation Research Part C: Emerging Technologies*, *90*, 156–165. https://doi.org/10.1016/j.trc.2018.03.005.
61. Zmud, J. (2018). Updating regional transportation planning and modeling tools to address impacts of connected and automated vehicles. *NCHRP report: Vol. 896*. Washington, DC: Transportation Research Board. https://doi.org/10.17226/25319.

Transportation Effects of CAD in Japan

Hiroaki Miyoshi, Shoji Watanabe, and Masanobu Kii

Abstract As part of the Cross-Ministerial Strategic Innovation Promotion Program of Automated Driving for Universal Services (SIP-adus) spearheaded by Japan's Cabinet Office, we construct microeconomics-based models to simulate the market diffusion of AVs, using the predictions of these models as a common source of basic numerical data to characterize the impact of AVs in various sectors. In this chapter, we briefly introduce our way of classifying automated driving systems, and two models developed to simulate the diffusion of AVs. We then present the results of a variety of variable-sensitivity analyses conducted using these models, and their policy implications. The main finding of the analyses is that enhancement of consumer expectations will be crucial for ensuring the diffusion of AVs in the future.

1 Introduction

Automotive technology and the automobile industry are in the midst of the kind of major revolution which occurs once in a century. Among the drivers of this upheaval are the four factors known collectively by the acronym **CASE**: "Connected" cars,

---

The original version of the chapter has been revised: Figures 5, 6, and 7 has been replaced with revised figures. A correction to this chapter can be found at
https://doi.org/10.1007/978-3-031-59876-0_9

H. Miyoshi (✉) · S. Watanabe
Doshisha University, Kyoto, Japan
e-mail: hmiyoshi@mail.doshisha.ac.jp

S. Watanabe
e-mail: showatan@mail.doshisha.ac.jp

M. Kii
Kagawa University, Takamatsu, Japan
e-mail: kii@see.eng.osaka-u.ac.jp

"Autonomous/Automated" driving, "Shared/Service", and "Electric".[1] Working together, these factors are poised to transform not just the automobile industry and the transport of people and things, but also the industrial structure of nations, the structure of cities, and the way we go about the day-to-day activities of our lives. As part of the **Cross-Ministerial Strategic Innovation Promotion Program of Automated Driving for Universal Services (SIP-adus)** spearheaded by Japan's Cabinet Office, the University of Tokyo and Doshisha University have recently led two research projects—*Study of Socioeconomic Impacts of Automated Driving Including Traffic Accident Reduction* (2018–2021), and *Research on Assessment of the Impact of Automated Driving on Society and the Economy and on Measures to Promote Deployment* (2021–2023). The objective is to quantitatively elucidate the extent to which automated vehicles (AVs) can help Japan solve relevant social issues: reducing traffic accidents and congestion, ensuring mobility for vulnerable road users, mitigating the driver shortage, and reducing the costs of logistics and mobility services (Cabinet Office [5]).

These research projects also seek to quantify how the adoption of AVs will influence industrial structure and productivity in Japan. To describe this influence in a consistent manner across several sectors, we construct micro-economics-based models to simulate the market diffusion of AVs, using the predictions of these models as a common source of basic numerical data to characterize the impact of AVs in various sectors. These simulation models also allow us to assess quantitatively the effectiveness of government policies to stimulate market diffusion, such as incentives to adopt AVs, or regulations mandating the installation of certain devices. In this chapter, we briefly outline the Japan-side simulation models developed for these purposes, then present the results of a variety of variable-sensitivity analyses conducted using these models, and their policy implications.[2]

2 Why Japan Needs Automated Vehicles

Before describing our simulation model, we first review several aspects of Japanese society that have made Japan particularly eager for a broad deployment of AVs.

Among the many ways in which AVs are expected to benefit society and improve daily life, the promise of fewer traffic accidents stands out as the most eagerly anticipated advantage of AVs in Japan. SIP-adus referred to traffic-accident reduction at

---

[1] The acronym CASE was used for the first time by Dr. Dieter Zetsche, *CEO* of *Daimler AG* and head of Mercedes-*Benz* Cars, at the 2016 Paris Motor Show.

[2] The University of Tokyo and Doshisha University are currently in the process of redesigning our Japan-side simulation models to reflect revisions including the classification of automated vehicles within our ongoing research program (*Research on Assessment of the Impact of Automated Driving on Society and the Economy and on Measures to Promote Deployment*). The model formulation and simulation results presented in this chapter pertain to the previous version of the model, and thus do not represent the final conclusions of the above-mentioned SIP-adus research projects or the official view of SIP-adus.

the top of its list of objectives for both Phase 1 (2014–2018) (Cabinet Office [4]) and Phase 2 (2018–2022) (Cabinet Office [5]) of the program. Annual traffic-accident fatalities in Japan peaked at 11,452 in 1992 and then began to decline, falling to 2,636, or 23% of the 1992 value by 2021. This decrease in fatalities was accompanied by a reduction in the total amount of economic losses associated with traffic accidents. Miyoshi [17] estimates the contributions of various individual factors to this reduction, one finding was that a substantial portion of the reduction could be attributed to technological advances in, and increased adoption of, *passive* safety provisions, such as airbags and safety body structure to absorb collision, which mitigate harm when accidents happen. However, the study also found that the impact of passive safety technologies was gradually decreasing, and concluded that further reductions in accident-related economic losses would depend critically on broader adoption of *active* safety technologies, which seek to prevent accidents from happening.

A second problem that Japan hopes to alleviate with AVs is a nationwide professional driver shortage (see chapters "Setting the Scene for Automated Mobility: A Comparative Introduction to the Mobility Systems in Germany and Japan, Governance, Policy and Regulation in the Field of Automated Driving: A Focus on Japan and Germany, and Social Acceptance of CAD in Japan and Germany: Conceptual Issues and Empirical Insights"). One way to understand this issue is to compare the age distribution of Japan's workforce to that of its professional drivers in Japan's 2015 national census [25], where persons aged 50 or above constituted 41.7% of all workers, but 56.6% of all professional drivers. Below age 50, younger and younger age segments account for smaller and smaller proportions of both workers and drivers, but the decrease is especially precipitous for professional drivers, suggesting that recruitment among younger workers has decreased. The retirement of a broad swath of Japan's workforce over the next 10 years is predicted to create a massive deficit of drivers; strategies to address this shortfall include traveling in a long platoon and pinning high hopes on the utilization of SAE Level-4 AVs.

Finally, we discuss the weakening of public transit networks in less-populated regions of Japan (see chapters "Setting the Scene for Automated Mobility: A Comparative Introduction to the Mobility Systems in Germany and Japan and Social Acceptance of CAD in Japan and Germany: Conceptual Issues and Empirical Insights"). Bus routes in rural areas are unprofitable and are being eliminated at a steady pace [19]. For older citizens no longer able to drive cars, this threatens to remove a source of mobility that is crucial for daily life (see e.g., Hanaoka [7]). The urgency of this problem has created high hopes that self-driving buses may furnish a solution, and trial experiments are currently underway in various regions around Japan.

3 Overview of Models to Simulate Market Diffusion of Automated Vehicles

We now present an overview of the simulation models developed to analyze the market diffusion of AVs.

3.1 Classification of Automated Vehicles

For the purposes of this model, we classify vehicles equipped with advanced driver-assistance systems or automated driving systems into 8 categories based on SAE driving automation-levels [23], on highways and on general roads. Automation on general roads will be more difficult to achieve than on highways. So, the timing for the introduction of SAE Levels 3 or 4 on general roads will be delayed compared to highways. It is for this reason that the types of roads are considered for classification. The classification scheme is shown in Table 1. Categories D0 through D3, corresponding to SAE Levels 1 and 2, are for vehicles equipped only with driver-assistance capabilities, while Categories A1 through A4 contain vehicles with more advanced capabilities.

We present two models, one dynamic model and one static model; the dynamic model is used to estimate the diffusion process of AVs up to SAE Level 4 (vehicles in categories D0-D2 and A1-A3), while the static model is used to estimate the spread of AVs that do not need a human driver (vehicles in category A4).

3.2 Dynamic and Static Simulation Models

A model frequently used to simulate the market diffusion of goods and services is the *S-shaped growth curve* (see e.g., [3, 6, 14, 21]. Rogers [22] classified consumers into 5 groups based on the rapidity with which they purchase new products: *innovators*, *early adopters*, *early majority*, *late majority*, and *laggards*. The adoption of new products begins with innovators, who have a deep interest in new ideas and are willing to take risks, and continues through the more conservative categories of consumers. Assuming a bell-curve distribution of consumers among the 5 categories, the cumulative distribution curve obtained by plotting product adoption versus time resembles the letter S; this is the S-shaped growth curve. S-shaped curve models have been used in several studies, including Litman [15], Mazur et al. [16], and Trommer et al. [27] to predict market-diffusion rates for AVs.

However, in our model we do *not* use the S-shaped growth curve, for two reasons. First, driver-assistance and automated driving systems are rather complicated when viewed as products. In the market for driving-assisted and self-driving vehicles, all 8 categories of vehicles in Table 1 are simultaneously present, and the benefit derived by

Table 1 Classification of vehicles equipped with advanced driver-assistance systems or automated driving systems

Category	Highways	General roads	Compatible technologies	Dynamic model	Static model
Advanced driver assistance system					
D0	SAE Lv. 1 or less	SAE Lv. 1 or less	Level under D1	✓	
D1	SAE Lv. 1 Driver assistance	SAE Lv. 1	Equipped with all the following four devices: • Collision-damage-reducing brakes, • Acceleration limiters for accidental accelerations (due to driver error), • Lane-departure warning system, and • Car distance warning system	✓	
D2	SAE Lv. 2 Partial automation	SAE Lv. 1	In addition to D1: • On highways, lane keeping systems (LKAS) + adaptive cruise control (ACC), and • Automatic lane changing on highways	✓	
D3	SAE Lv. 2 Partial automation	SAE Lv. 2	In addition to D2: • Lv. 2 on general roads		✓
Automated driving system					
A1	SAE Lv. 3 Conditional automation	SAE Lv. 2	In addition to D3: • Lv. 3 on highways, and • Lv. 2 on general roads	✓	
A2	SAE Lv. 4 High automation	SAE Lv. 3 on major arteries and thoroughfares	In addition to A1: • Lv. 4 on highways, • Lv. 3 on major general roads, and • On general roads, take-over requests (TORs) for driving operations will be issued in response to system demand	✓	
A3	SAE Lv. 4 High automation	SAE Lv. 4 on major arteries and thoroughfares	In addition to A2: • Lv. 4 on major general roads, and • Take-over requests (TORs) will not be issued	✓	
A4	Driveress vehicle equivalent to SAE Lv. 4 or 5				✓

Source Based on Table 1 in [26] (p. 137), with revised category names and other minor changes a user of vehicles with these systems varies greatly depending on factors such as the transportation environment near the user's home and the user's own driving ability, and there is no guarantee that users will prefer systems in upper-tier categories. The market diffusion of goods like this is most accurately described by suitable adaptations of models emphasizing the utility enjoyed by individual consumers. Second, S-shaped growth-curve models, which are premised on the assumption of vigorous communication and interaction among various constituents of societies, are poorly suited to the analysis of factors such as price change of automation systems and government incentives to promote their adoption; again, the influence of such phenomena is best captured by models based on individual utility, which treat prices and similar factors as explanatory variables.

For these reasons, we used web-based online questionnaires to gather data on stated consumer preferences, then inferred consumer demand based on this data, to construct two models, a dynamic model and a static model, in which market diffusion is driven by the relationship between supply pricing and consumer demand. The reason that we constructed two models is that the timing for the introduction of AVs up to category A3 is already being discussed and can be predicted with reasonable confidence, while the restrictions on types of roads enabling driverless vehicles (category A4) makes their feasibility and the attendant timing impossible to predict. In the classification scheme of Table 1, the dynamic model covers the driver-assistance systems of categories D0-D2 and the automated driving systems of categories A1-A3. We simulate market-diffusion rates for products in each of these categories in the form of time series extending to 2050. The static model encompasses categories D3, the most advanced driving-assistance systems, and A4, automated driving systems requiring no driver; the model assumes that these systems—and only these systems—are simultaneously available, and simulates the market diffusion of vehicles with each type of system under steady-state conditions. These models are described in detail below.

(1) Dynamic model

The dynamic model considers three purposes of vehicles: privately-owned vehicles (passenger cars), commercial transport vehicles (trucks), and public-transit vehicles (buses and taxis). For vehicles with each purpose and driving systems in categories D0-D2 and A1-A3, the model predicts the number of vehicles, and the total travel distance for each category of system from 2015 to 2050. The prediction methods differ for vehicles with different purposes; here we describe the method used to simulate the diffusion of privately-owned AVs, following the discussion in [26].

For these simulations, we separately consider three sizes of passenger cars: standard-sized cars,[3] small-sized cars, and mini cars (Japanese "kei"-cars). We first use predictions for the population of citizens with driver's licenses and for per-capita GDP to estimate the total number of vehicles (of all three sizes) owned in various years in the future, then use a distribution formula to subdivide this total among

---

[3] For example, gasoline *cars* with the *engine displacement* of above *2000 cc are included in 'standard sized cars'.*

the three sizes to calculate the number of owned vehicles of each size that will be required in each year. Next, we estimate the number of vehicles scrapped in a given year by segregating vehicles owned in the previous year by vehicle age and multiplying by the rates of elimination, which are assumed to follow the actual value and/ or the Weibull distribution (see e.g., Kilde and Lasen [12]); subtracting the number of discarded vehicles from the number of owned vehicles in the previous year yields the number of retained vehicles in the given year. We repeat this calculation for all categories of driving-assistance and AVs, assuming that rates of elimination are the same for both conventional vehicles and AVs.

For a given year, subtracting the number of retained vehicles from the required number of owned vehicles yields the total number of *new* vehicles to be acquired that year. The distribution of this total number among the various categories of driving-assistance and automated driving systems is determined by the relationship between supply and demand, as indicated in Fig. 1. The purple curves on the left in Fig. 1 indicate *acceptance rates* for each system category; we determine the acceptance rate for each system—defined as the fraction of all consumers willing to purchase the system at a given price- -based on consumer preference data obtained from the online questionnaires. In the questionnaires, respondents were given a description of one category of driving-assistance or automated driving system and asked, in a double-bounded dichotomous choice, to state their willingness to purchase the system at randomly chosen prices.[4] Note that the explanatory variables used to determine acceptance-rate curves include not only the prices of driving-assistance and automated driving systems, but also additional factors such as respondents' expectations for automated driving, as described later.

On the supply side, shown on the right in Fig. 1, system prices for each year reflect the impact of mechanisms governed by *experience-curve effects*; phenomena, observed for many types of products (e.g., Haysom et al. [8], Latimer and Meier [13], Nykvist and Nilsson [20], Schmidt et al. [24]), in which increases in aggregate production volumes spur improvements in manufacturing technology, reducing product costs. Here we assume that experience-curve effects apply to AV technologies. For each category of driving-assistance or automated driving system, the acceptance rate—that is, the proportion of consumers willing to purchase the system—is determined by the intersection of the consumer acceptance rate and price curves. For this model, we assume that acceptance rates are determined based on acceptance rate curves and system prices in descending order of system category (i.e., for categories A3, A2, A1, and D2, in that order) and that the proportion of all new vehicles represented by a category is given by the difference between the acceptance rates for that category and the next highest category. Below category D2, new vehicles are proportionally allocated to categories D1 and D0 based on exogenously determined

---

[4] Online surveys were conducted in Japan in December 2019. All respondents were current or former owners of personal passenger vehicles and current or former holders of driver's licenses. Respondents were asked to consider replacing their own vehicle with a newly-purchased vehicle of the same type. We received a total of 6,156 questionnaire responses, from which we excluded responses with incorrect answers to questions testing understanding of explanatory text passages. This yielded a set of 1,828 responses from which we determined acceptance-rate curves.

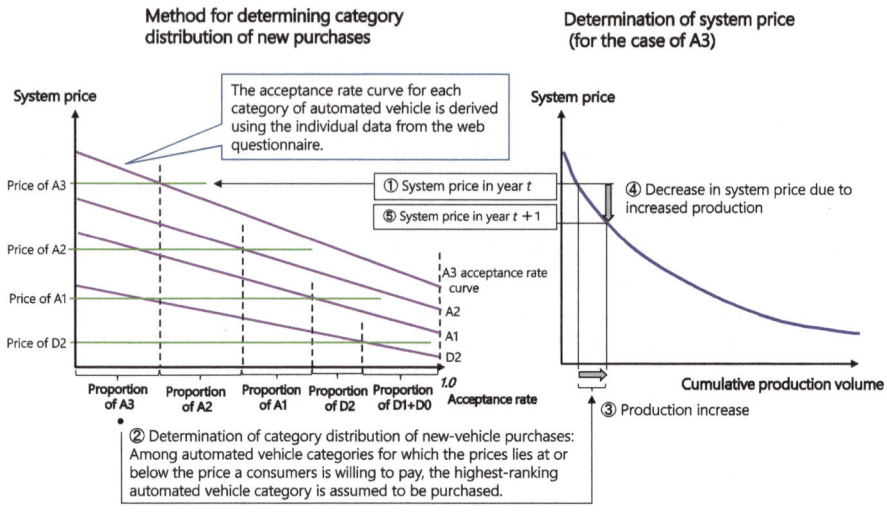

Fig. 1 Dynamic model concept. *Source* Based on Fig. 2 in Suda and Miyoshi [26] (p. 138)

ratios. Note that our calculations of total travel distance assume that per-vehicle travel distances for vehicles of each type remain unchanged from their 2015 values; in other words, this model does not account for the possibility that the advent of AVs may affect consumers' transport mode preferences or travel distances. As discussed below, this is a significant point of difference between our dynamic and static models.

(2)   Static model

If and when new forms of transportation services—such as driverless passenger cars, driverless taxis based on such vehicles, or shared-ride driverless taxis—become available, consumer choices regarding modes of transportation may shift significantly compared to the 2022 status quo. The objective of our static model is to simulate these variations in consumer choices and their impact on vehicle ownership and travel distances. This static model considers only travel by passenger cars, excluding other types of vehicles. Also, the model considers only two of the eight categories of technology system in Table 1: D3, encompassing vehicles that can use SAE Level-2 functionality on both highways and general roads, and A4, corresponding to driverless AVs. Consumers are assumed to choose from six available transport modes: (1) privately-owned manual cars (category D3), (2) privately-owned driverless cars (A4, one passenger), (3) driverless taxis (A4), (4) shared-ride driverless taxis (A4), (5) public transit (bus and/or rail), and (6) walking or bicycling. As illustrated in Fig. 2, our static model is comprised of three sub-models: (I) Consumer mode choice model, (II) a model to simulate numbers of owned vehicles and travel distances for privately-owned passenger cars, and (III) a model to simulate the fleet size and travel distances for passenger cars used to provide driverless taxi services.

(I)   *Consumer mode choice model*

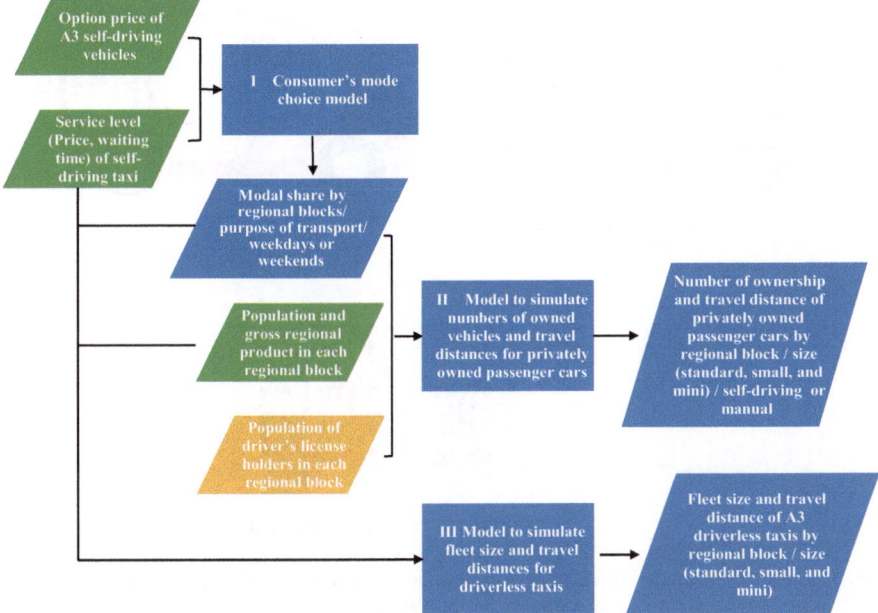

Processing or data in this particular simulation model

Exogeneous variable

Processing or data of other simulation model in this Project.

Fig. 2 Static model concept

Assuming a hypothetical scenario in which driverless passenger cars are released to market and consumers are able to own them privately or use transportation services based on them, this model estimates the probability with which consumers select each of the six transport modes and the number of trips completed via each mode. Estimates are generated separately for trips of various types under various conditions: trips on weekdays vs. weekends and holidays, trips of various purposes and various travel distances (1, 3, 10, and 30 km) and trips in large cities vs. non-large cities. Here "large cities" refers to Japan's three major metropolitan areas (Tokyo, Nagoya, and Osaka) and cities designated by government ordinance. Transport mode choices are formulated based on stated consumer preference data obtained from online questionnaires. In these questionnaires, each respondent was asked to select one of five possible objectives for daily travel: commuting to work, commuting to school, shopping, medical appointments, or other personal business. A travel distance appropriate for the selected objective was then chosen at random, and respondents were informed of the cost and time required to complete a trip of this distance via each transport mode, and asked which mode they would choose for such a trip. To have respondents depict long-term mobility decisions, respondents were asked *not* to consider

a)   **Privately-owned driverless car**
(Equipped with fixed mirror)

b) Driverless taxi

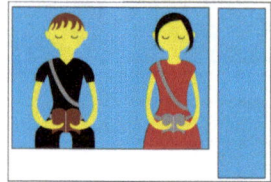

c)   **Shared- ride driverless taxi**

Fig. 3 Examples of images of vehicle interiors shown to online-questionnaire respondents to help create accurate mental images of technologies that are not currently available

the typical transportation conditions of their actual lives, but rather to respond as they would upon moving to a new home and beginning a new lifestyle. To assist in developing accurate mental images of the three currently non-existent transport modes (privately-owned driverless cars, driverless taxis, and shared-ride driverless taxis), questionnaire respondents were presented with several illustrations like those shown in Fig. 3.[5]

Our model for predicting transport mode choices is a multinomial logit model whose explanatory variables include the price for the A4, level-of-service conditions (fares and wait times) for driverless taxis (including shared-ride driverless taxis), wait times for public transit networks, and the expectations of respondents regarding driverless vehicles. Note that this model encompasses three transport modes that do not currently exist: privately-owned driverless cars, driverless taxis, and shared-ride driverless taxis. To tune the model, we set prohibitively high prices for these three modes, then adjusted the model until the consumer choice probabilities it predicts for the remaining three modes (privately-owned manual cars, public transit, and walking or bicycling) match the estimated choice probabilities in 2015 for cities of five size classes.[6] In other words, the adjustable model parameters account for the

---

[5] Online questionnaires were administered to respondents aged 20 or above in three phases between October 8 and December 27, 2020. After studying consumer responses received in phases 1 and 2, we revised some aspects of the various transportation modes and their descriptions in questionnaires, then administered phase 3 as the main survey, with 5,102 respondents.

[6] The 2015 choice probabilities are estimated using *the Nationwide Person Trip Survey* provided by the Ministry of Land, Infrastructure, Transport and Tourism.

dependence of various factors—such as the quality of public-transit services and parking services—of each city size. Then, the adjusted coefficients determined in this way for privately-owned manual cars and public transit were applied respectively to privately-owned driverless cars and driverless taxis. Using the multinomial logit model obtained in this way, various types of data, such as prices for driverless AVs and service qualities (price and wait time) for driverless taxis (including shared-ride taxis) were input. Then, consumer choice probabilities and trip counts for each transport mode under various conditions, such as travel on weekdays versus weekends or holidays, and trips of various purposes and various travel distances can be calculated.

(II) *Model to simulate numbers of owned vehicles and travel distances for privately-owned passenger cars*

For privately-owned passenger cars, we used 2010 and 2015 panel data for 15 geographical regions to formulate a relationship describing how per-capita numbers of owned vehicles and travel distances vary with factors such as the modal share of all trips made by privately-owned passenger cars segregated by trip purpose, trip timing (weekdays vs. weekends and holidays), per-capita GRP, and the region's population density. Using the proportion of privately-owned vehicles equipped with systems in categories D3 and A4 (as computed by the consumer mode choice model described above), and populations of geographical blocks [18], this model computes, for each geographical block, numbers of privately-owned vehicles owned and total travel distances.

(III) *Model to simulate fleet size and travel distances for driverless taxis*

For driverless taxis, we first formulated a relationship between the number of trips of various distances and actual travel-distance demand. Then we input the number of trips of various distances taken by driverless taxis (including shared-ride driverless taxis, as computed by the consumer choice model described above) and computed the actual travel-distance demand for driverless taxis in each geographical block. Combining this actual travel-distance demand with the distance traveled without paying passengers yields the total travel distance. The distance traveled without paying passengers includes dead mileage estimated from active and dead mileage rates set by referencing to previous works (e.g., Azuma et al. [1, 2], Kamijo et al. [9], Katsuki et al. [10]). The number of owned vehicles is computed as the number of vehicles required to ensure expected service-quality levels (in particular, mean taxi wait times) at times of maximal utilization of driverless taxis, by using the methodology in Kii et al. [11].

3.3 *Assumptions*

For the purposes of these simulations, we assumed that the initial market releases of the various categories of driving-assistance and automated driving systems will take

Table 2 Years and prices of initial market releases of driving-assistance and automated driving systems

	Initial market releases	Prices (JPY)
D2	2025	150,000
A1	2030	400,000
A2	2040	450,000
A3	2045	450,000
A4	–	1,000,000

place at the years and prices shown in Table 2. These prices represent an additional cost on top of conventional vehicle prices, based on various materials.

4 Results from Dynamic Model

We now discuss the results of simulations using the models described above, beginning with the dynamic model.

Figure 4 shows the model's predictions for 2030, 2040, and 2050 of the proportions of passenger cars (for both private and commercial use) equipped with the various categories of driving-assistance or automated driving systems. Considering categories A1 and above, representing automated driving vehicles, we see that only 1.9% of vehicles are predicted to be equipped with these systems by 2030, but this grows to 23.7% by 2040, and to 45.9% by 2050 (cf. Fig. 7 in chapter "Transportation Effects of Connected and Automated Driving in Germany"). This is explained by noting that systems in categories A1, A2, and A3 are released to market in 2030, 2040, and 2045, with the increase in cumulative production of systems in each category creating experience-curve effects that reduce supply prices.

Meanwhile, for categories D2 and below, corresponding to driving-assistance systems, we see that by 2030 the proportion of vehicles equipped with systems in category D0 falls to just 21.3%. This is because our simulation system assumes that the installation of category-D1 technologies will become mandatory in 2022. In view of the fact that driving-assistance systems can help to avoid the majority of traffic accidents, this suggests that—from the perspective of reducing traffic accidents—relatively large benefits will be achieved by 2030.

We next analyze how variations in initial retail prices, and in consumer expectations regarding AVs, affect the market diffusion of AVs with driving-assistance systems or automated driving systems. For this purpose we consider four scenarios: (1) a 50% decrease in initial retail prices (Scenario 1), (2) an increase in consumer expectations regarding the benefit they themselves will derive from automated driving (Scenario 2), (3) an increase in consumer expectations regarding the benefit to society from automated driving in addition to Scenario 2 (Scenario 3), and (4) Scenarios 1, and 3 simultaneously (Scenario 4).

Consumer expectations in Scenarios 2 and 3 are defined as follows. The online questionnaires asked respondents to rate their expectations, on a scale from 1 (no

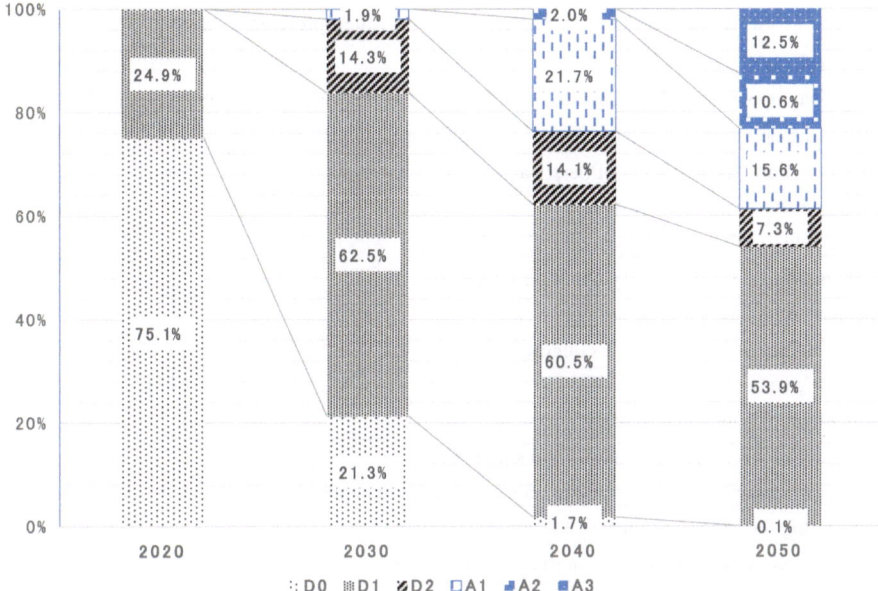

Fig. 4 Percentage breakdown of driving-assistance and automated driving systems in passenger cars. *Note* The 2020 values are estimated

expectations) to 5 (high expectations), of 13 potential benefits of automated driving, including six benefits to individuals and seven benefits to society, as shown in Table 3. The total expectation scores for the six personal benefits and the seven societal benefits, averaged over all respondents, were 20.5 and 25.0 respectively. Scenarios 2 and 3 simply assume that these averages take their *80th percentile values* (24.0 and 30.0 respectively).

Figure 5 shows how adoption of the various categories of driver-assistance or automated driving systems among privately-owned cars (of all size classes) in 2030, 2040, and 2050 varies under the four scenarios. We see that each of the three scenarios 1, 2, and 3 has the effect of accelerating the adoption of the most advanced self-driving systems available on the market at a given time. What is interesting here is that Scenario 3, raising consumer expectations regarding the benefits of self-driving cars—both to themselves and to society—has greater efficacy than a 50% reduction in initial market prices in stimulating the adoption of more advanced self-driving systems. This finding indicates that education and outreach efforts to ensure thorough societal understanding of the benefits of self-driving vehicles will be a key priority for ensuring their spread.

Table 3 Potential benefits of automated driving in the stated-preference surveys

Benefits to individuals
Increasing opportunities for trips like shopping, leisure activities, and hobby projects
Increasing opportunities to visit friends, acquaintances, family, and relatives
Reducing burdens on drivers of personal vehicles
Making use of time spent traveling
Being able to call vehicles from other locations
Eliminating the need to worry about parking during trips
Benefits to society
Reducing or eliminating traffic congestion
Reducing traffic accidents
Reducing environmental burden
Assisting mobility for older adults and some others
Alternative transportation for public transit networks in depopulated regions
Alleviating the shortage of truck, bus, and taxi drivers
Spurring economic activity and international competitiveness

Source: These items in the online survey were provided by Prof. Ayako Taniguchi, University of Tsukuba

5 Results from Static Model

In this section, we present simulation results from our static model, which considers scenarios for the market diffusion of driverless vehicles. Here, we choose 2015 as a reference year and consider, as our reference scenario, a hypothetical scenario in which driverless AVs are available in that year. In addition to this reference scenario, we consider six variant scenarios listed in Table 4: the price of driverless vehicles decreases (Scenario 1), the price of driverless taxis (including shared-ride driverless taxis) decreases (Scenario 2), the wait time for driverless taxis (including shared-ride driverless taxis) decreases (Scenario 3), consumer expectations for both personal and social benefits increase (Scenario 4), and wait times for public transit networks decrease (Scenario 5), and scenarios 1, 2, and 4 occur simultaneously (Scenario 6). For each of these variant scenarios, we consider how consumer transport mode choices vary relative to the reference scenario.

Figure 6 plots, for geographical blocks containing Japan's three major metropolitan areas (Tokyo, Osaka, and Nagoya), the modal share of the various transport modes (a), total daily travel distances for trips by automobile (b), and numbers of passenger cars owned (c). We first discuss results for the reference scenario, the year 2015 with driverless vehicles available on the market. In this scenario, privately-owned manual cars accounts for 28.6% of all trips; privately-owned driverless cars, 12.1%; driverless taxis, 7.2%; shared-ride driverless taxis, 6.2%; public transit, 17.4%, and walking or bicycling, 28.5%. How do these results compare to 2015 values? The 2015 values plotted in the graph represent the share of all trips made by motor vehicles (including corporate-owned vehicles and taxi), by public transit

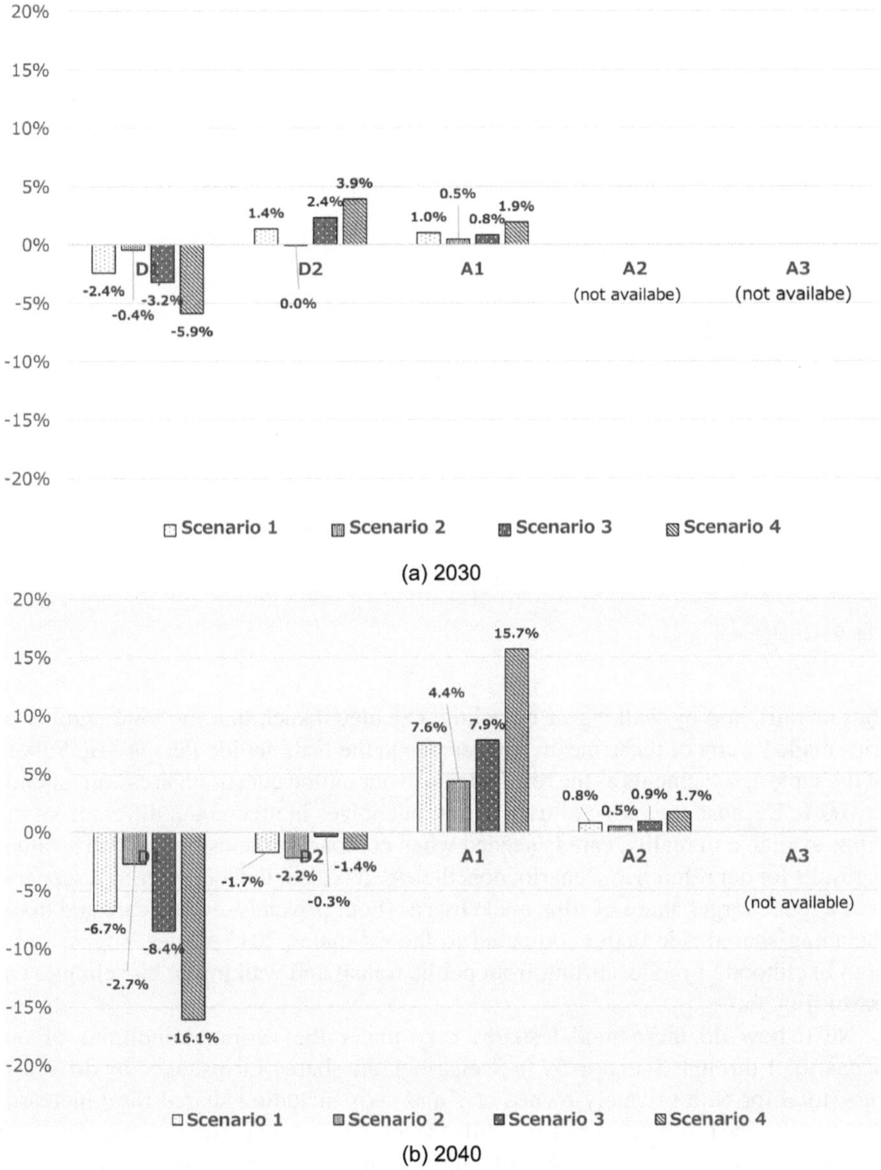

Fig. 5 Results of dynamic-model sensitivity analysis

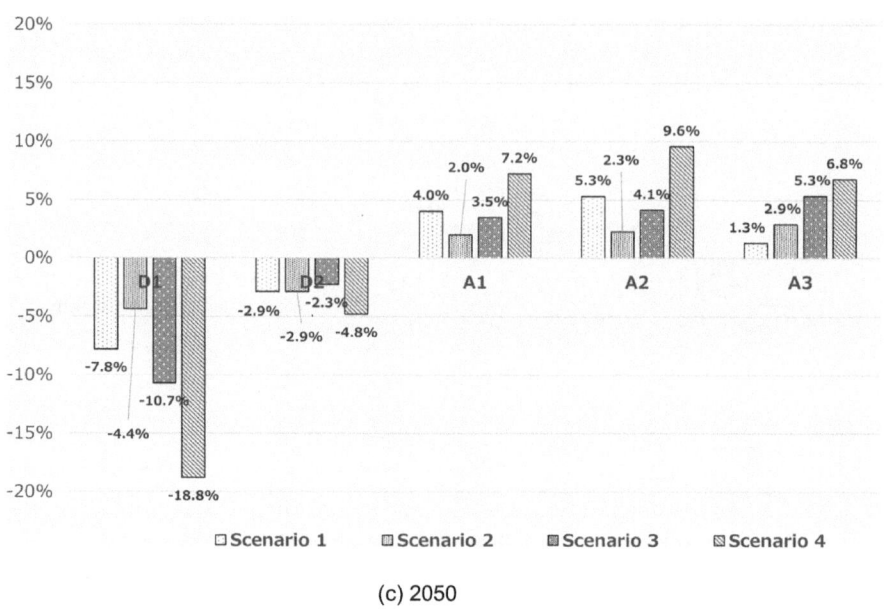

(c) 2050

Fig. 5 (continued)

(bus or rail), and by walking or bicycling, calculated such that the total number of trips made by any of these means, by persons in the Nationwide Person Trip Survey of the same age segments as the respondents to our online questionnaires corresponds to 100%. Because the range of transportation choices in this model differs from the range available in reality, care is needed when comparing the estimated 2015 values to results for our reference scenario; nonetheless, it is clear that our reference scenario sees a much larger share of trips made by car (both privately-owned cars and taxis, including shared-ride taxis) compared to the estimated 2015 values, suggesting a high likelihood of mode shifting from public transit and walking or bicycling to car travel (Fig. 6a).

Next, how do these modal shares vary under the various conditions of our Scenarios 1 through Scenario 5? In Scenario 4, the share of trips taken by driverless cars (total for both privately-owned cars and taxis including shared-ride) increases compared to the reference scenario, while the modal share of privately-owned manual cars decreases together with that of public transit. The share for driverless taxis (including shared-ride taxis) also increases in Scenarios 2 and 3, but not as much as in Scenario 4. Also, in Scenario 1 the share for privately-owned driverless cars remains largely unchanged. In this study, we do not use elasticity to relativize degrees of influence, and thus we cannot rigorously compare sensitivities to financial factors, such as the price of driverless technologies or driverless taxis, against sensitivities to consumer expectations; nonetheless, this result suggests that higher consumer expectations for automated driving may be more influential than vehicle prices in promoting adoption of driverless vehicles. We explain this by noting that many respondents remain skeptical of the benefits of driverless cars and do not consider

Table 4 Scenarios used for static-model sensitivity analysis

Scenario		Description
2015		Driverless vehicles, driverless taxis, and shared-ride driverless taxis not yet available
Reference scenario		System price for driverless vehicles: 1 million JPY; driverless taxi fare: 25 yen/km; average wait time for driverless taxis: 15 min; consumer's average expectation score: 45.5; average wait time for public-transit networks: 17 min
Scenario 1	Price reduction for driverless vehicles	Price of driverless driving capabilities reduced from 1 million JPY to 0.5 million JPY
Scenario 2	Price reduction for driverless taxis	Driverless taxi fare reduced from 25 yen/km to 15 yen/km (prices for shared-ride driverless taxis are set at 1/1.3, where 1.3 is the average number of passengers)
Scenario 3	Decrease in wait time for driverless taxis (including shared-ride driverless taxis)	Average wait time decreased from 15 to 10 min
Scenario 4	An Increase in consumer expectations for both personal and social benefits	Total expectation scores for the six personal benefits and the seven societal benefits increased from 45.5 to 52.0
Scenario 5	Reduction in wait time for public transit	Average wait time for public-transit networks reduced from 17 to 6 min
Scenario 6		Scenario 1, Scenario 2, and Scenario 4 occur simultaneously

price to be the most important consideration. However, consumers who *do* understand the advantages of driverless vehicles exhibit strong preferences for these vehicles; moreover, when both privately-owned driverless cars and driverless taxis are available, a significant number of consumers shift from using private vehicles to using driverless taxis.

Next, Fig. 6b shows total daily travel distances for trips by car. Here, again, comparing actual 2015 statistics to values for our reference scenario shows that total daily travel distances in the reference scenario are significantly greater than actual 2015 statistics, suggesting that the introduction of driverless cars is highly likely to increase daily travel distances. Looking at Scenarios 1 through 5, we note again that significant deviations from the reference scenario are seen in Scenarios 4, in which automobile travel distances increase over the reference scenario by 13.5%.

Finally, in Fig. 6c for the number of owned vehicles, we see that the greatest deviations from the reference scenario in the total number of privately-owned driverless cars, driverless taxis, and shared-ride driverless taxis from the reference scenario occurs in Scenarios 2, 3 and 4. Note that the increase in Scenarios 3 includes total travel distances by driverless taxis both with and without paying passengers.

Next, Fig. 7 is similar to Fig. 6, but represents geographical blocks *not* containing the Tokyo, Osaka, or Nagoya areas. In the reference scenario, privately-owned manual cars account for 45.2% of all trips; privately-owned driverless cars, 18.2%;

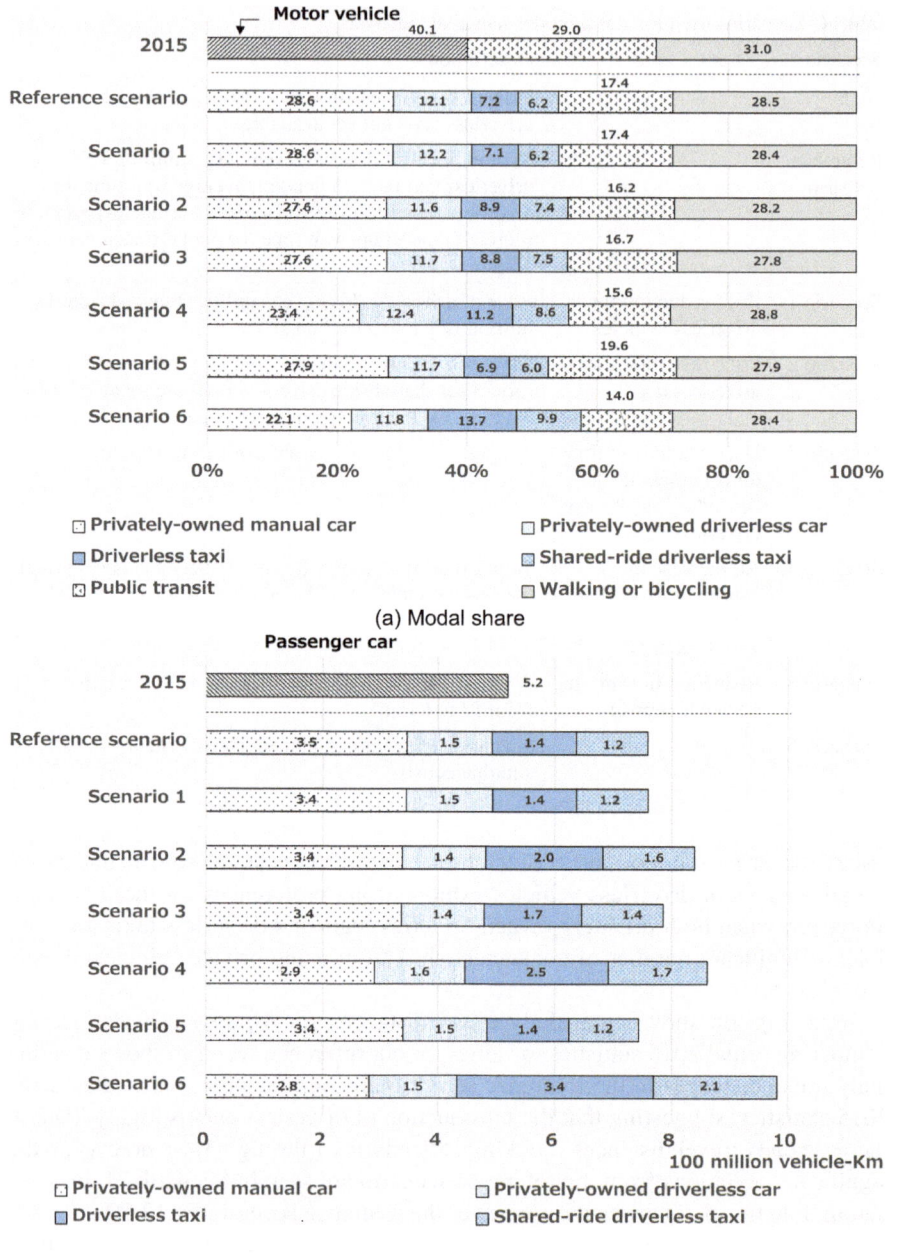

(a) Modal share

(b) Total daily travel distance

Fig. 6 Results of static-model simulations for geographical blocks containing the three major metropolitan areas Note: The 2015 values of (a) are calculated based on the data from the Nationwide Person Trip Survey provided by the Ministry of Land, Infrastructure, Transport and Tourism.The 2015 values of (b) and (c) are determined based on data from the Road Traffic Census provided by the Ministry of Land,Infrastructure, Transport and Tourism, and from the Automobile Inspection & Registration Association, respectively

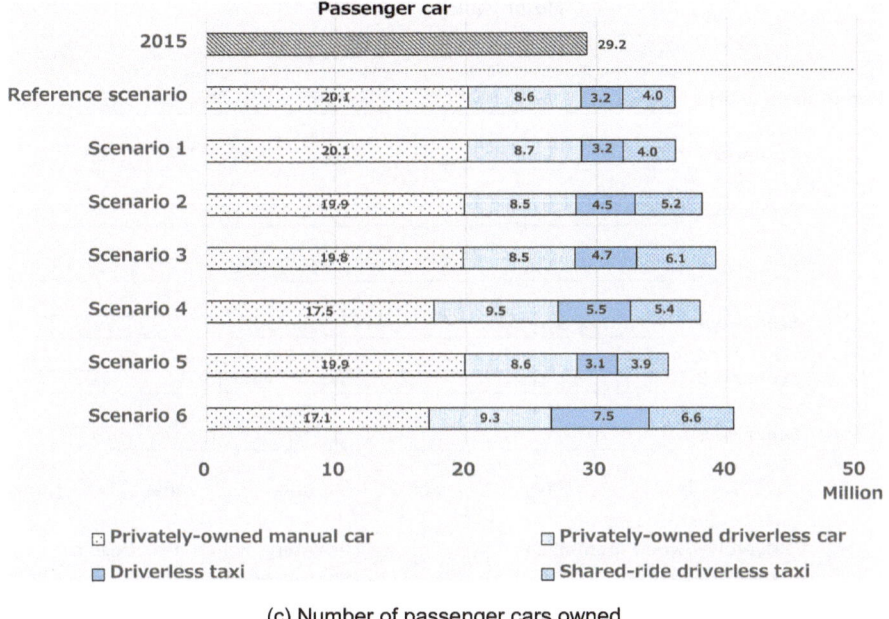

(c) Number of passenger cars owned

Fig. 6 (continued)

driverless taxis, 5.5%; shared-ride driverless taxis, 3.4%; public transit, 6.6%, and walking or bicycling, 21.1%. Compared to the results of Fig. 6 for geographical blocks containing the major metropolitan areas, we see greater modal share for cars in both 2015 data and the reference scenario (Fig. 7a). Next, comparing the estimated 2015 values to values for our reference scenario, we see that the share of trips made by car (total for both privately-owned cars and taxis, including shared-ride taxis) is higher in the reference scenario than in the estimated 2015 values, although the increase is not as large here as it was for Japan's three major metropolitan areas.

Results for Scenarios 1 through 5 are similar to the results discussed earlier for geographical blocks containing the major metropolitan areas; of the five scenarios, prominent differences in modal shares from the reference scenario are observed in Scenarios 4. We see that the total share of driverless taxis and shared-ride driverless taxis increases from 8.9 to 13.8%, while that of privately-owned driverless cars increase from 18.2 to 19.5%.

Next, in Fig. 7b we see that daily travel distances in our reference scenario are 14.3% greater than the estimated 2015 values, suggesting that the introduction of driverless cars may significantly increase travel distances. Results for Scenarios 1–5 again show significant differences for Scenarios 2 and 4, in which total daily automobile travel distances increase, as was true in Fig. 6b. From Fig. 7c we see that the total number of privately-owned driverless cars, driverless taxis, and shared-ride driverless taxis increases significantly for Scenarios 4.

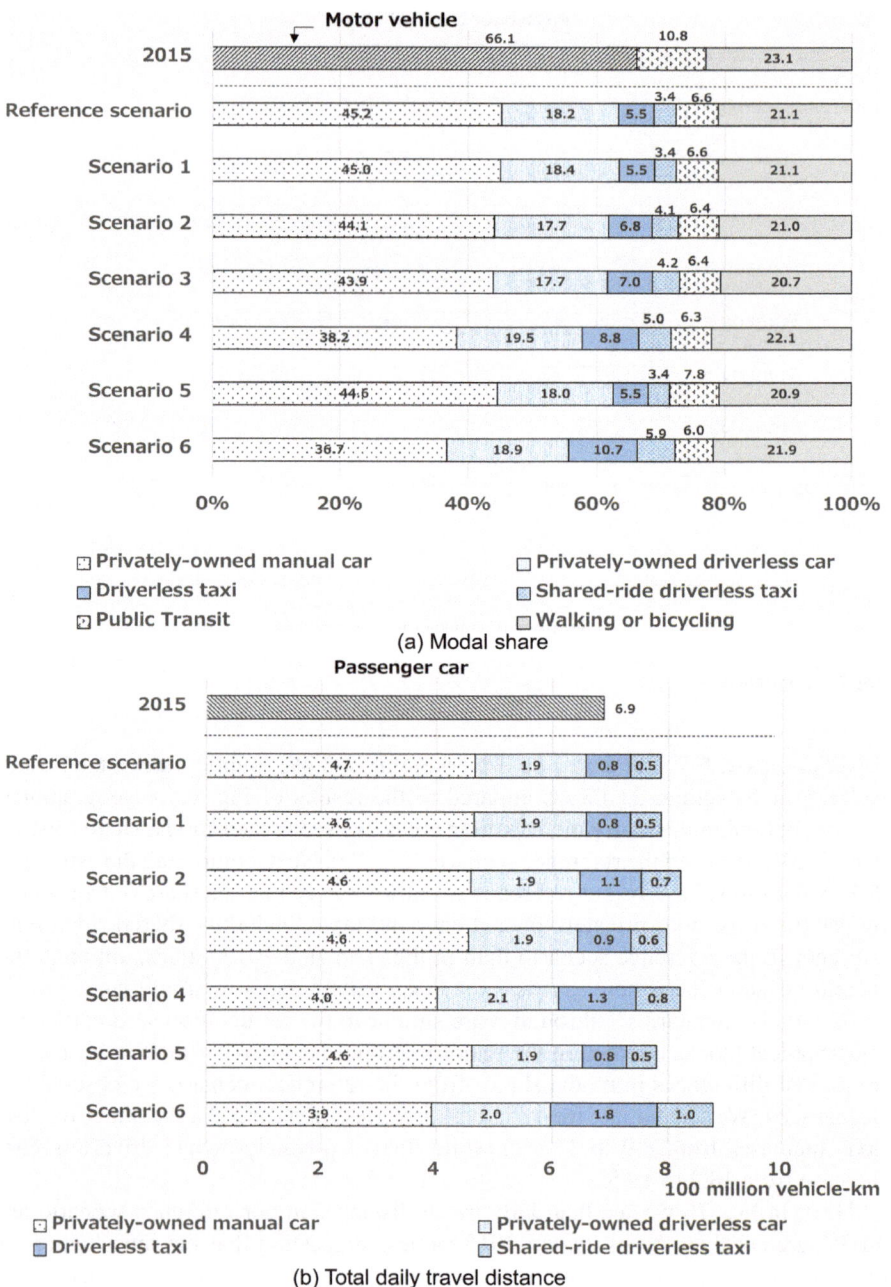

Fig. 7 Results of static-model simulations for geographical blocks *not* containing the three major metropolitan areas Note: The 2015 values of (a) are calculated based on the data from the Nationwide Person Trip Survey provided by the Ministry of Land, Infrastructure, Transport and Tourism. The 2015 values of (b) and (c) are determined based on data from the Road Traffic Census provided by the Ministry of Land,Infrastructure, Transport and Tourism, and from the Automobile Inspection & Registration Association, respectively

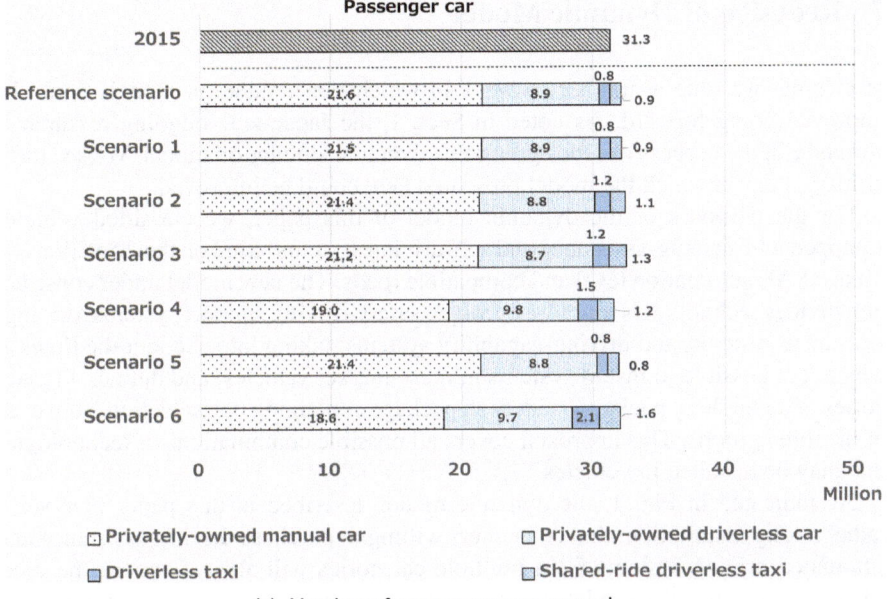

(c) Number of passenger cars owned

Fig. 7 (continued)

6 Conclusions

In this work, we formulated and discussed the predictions of two models for simulating market diffusion of AVs. The results of our simulations have two key implications.

First, appropriate enhancement of consumer expectations will be crucial for ensuring the spread of AVs in the future. The results of our dynamic model indicate that the adoption of more advanced categories of automated driving systems can be effectively stimulated by heightening consumers' expectations of the benefits. Also, our static model showed that higher expectations are more effective than lower costs in promoting adoption of the novel transportation paradigm of driverless taxis.

Second, as is clear from the simulation results of our static model, enhanced expectations regarding AVs together with reduced prices are likely to increase total automobile travel distances by inducing a modal shift toward driverless taxis. This may lead to higher energy-consumption and more traffic congestion, as well as increased demand for government spending in areas such as road infrastructure. However, driverless vehicles may bring benefits to society which are out of scope of this research. Those benefits may, for example, include the reduction of parking lots in city centers. Further detailed simulation models will be needed to evaluate such benefits.

7 Redesign of Dynamic Model

In closing, we note a number of ways in which our simulation models could be improved going forward. As noted in Sect. 1, the Japan-side ongoing research is currently in the process of redesigning this dynamic simulation model. We see three primary areas in which the model presented here could be improved.

For the purposes of the dynamic model of this paper, we classified vehicles equipped with driving-assistance and self-driving systems based on their positions on 2 axes: SAE automation level and compatible roads. The new model under construction divides technologies into three groups: safety-driving support systems, driving-assistance systems, and driving-capability systems; taking into account the times at which it is possible to install systems in mass-market vehicles, and defines 11 categories of technology package, with each package containing one selection from each of the three groups. This approach covers all possible combinations of technologies that may be installed in vehicles.

As indicated in Fig. 1, the dynamic model described in this paper involves a rather strong assumption: that consumers willing to purchase driving-assistance and automated driving systems from multiple categories will always choose the most advanced category. The model under construction introduces a more natural assumption: that consumers simply select one of the 11 system categories as their preference. We are currently conducting new consumer stated-preference surveys to assist in constructing this model.

The simulation models presented here account only for the *positive* aspects of consumer expectations regarding AVs, namely, expectations of benefits to consumers themselves and benefits to society at large. We are currently preparing new consumer stated-preference surveys which will consider also *negative* consequences of the spread of AVs, such as increased demand for government spending, and other factors associated with cost increases, to assess consumer acceptance. Incorporating the relationship between societal acceptance and market diffusion into the simulation model will allow us to assess the extent to which effective cultivation of societal acceptance can stimulate the spread of AVs.

Acknowledgements This paper is based on results obtained from a project, JPNP18012, commissioned by the New Energy and Industrial Technology Development Organization (NEDO).

Our research was greatly facilitated by assistance from Professor Makoto Itoh of University of Tsukuba regarding the functions of various categories of advanced driver-assistance systems and automated driving systems and their proper descriptions in questionnaires. In designing our online questionnaires, we received invaluable advice from Professor Ayako Taniguchi of University of Tsukuba on questions probing knowledge and expectations regarding AVs. We also received the gracious permission of Professor Taniguchi to use, with partial modifications, questionnaires she had previously prepared. Finally, Japan's Ministry of Land, Infrastructure, Transport and Tourism (MLIT) provided survey data from their "Nationwide Person Trip Survey" for 2010 and 2015, which we used in constructing the static model. We take this opportunity to extend our gratitude to Professor Ito, Professor Taniguchi, and MLIT.

References

1. Azuma, K., Katsuki, H., and Taniguchi, M. (2018), Urban structure influences on shared mobility–automated driving efficiency, Journal of the City Planning Institute of Japan, 53(3), pp. 551–557, https://doi.org/10.11361/journalcpij.53.551 (In Japanese).
2. Azuma, K., Katsuki, H., and Taniguchi, M. (2019), Urban structure influences on efficiency of shared mobility with automated driving, Transport Policy Studies' Review, 21, pp. 027–038, https://doi.org/10.24639/tpsr.TPSR_21R_04 (In Japanese).
3. Baur, L., and Uriona, M. M. (2018), Diffusion of photovoltaic technology in Germany: A sustainable success or an illusion driven by guaranteed feed-in tariffs?, Energy, 150, pp. 289–298, https://doi.org/10.1016/j.energy.2018.02.104.
4. Cabinet Office (2018), Cross-ministerial Strategic Innovation Promotion Program (SIP), Automated Driving System R&D Plan, April 1, 2018.
5. Cabinet Office (2021), Cross-ministerial Strategic Innovation Promotion Program (SIP), Automated Driving for Universal Services R&D Plan, May 26, 2021.
6. Fluchs, S. (2020), The diffusion of electric mobility in the European Union and beyond, Transportation Research Part D: Transport and Environment, 86, Article 102462, https://doi.org/10.1016/j.trd.2020.102462.
7. Hanaoka, T., Oguchi, T., and Toriumi, A. (2020), Establishment of social deployment requirements of automated driving systems, Seisan Kenkyu, 72(2), pp. 159–164, https://doi.org/10.11188/seisankenkyu.72.159 (In Japanese).
8. Haysom, J. E., Jafarieh, O., Anis, H., Hinzer, K., and Wright, D. (2014), Learning curve analysis of concentrated photovoltaic systems, Progress in Photovoltaics: Research and Applications, 23(11), pp. 1678–1686, https://doi.org/10.1002/pip.2567.
9. Kamijo, Y., Luo, L., Parady, G. T., Takami, K., and Harata, N. (2019), Scenario evaluation of autonomous vehicle spread using agent-based simulation, JSTE Journal of Traffic Engineers, 5(2) (Special Edition A), pp. A_142–A_151, https://doi.org/10.14954/jste.5.2_A_142 (In Japanese).
10. Katsuki, H., Azuma, K., Takahara, I., and Taniguchi, M. (2018), Change of vehicle utilization by shared mobility with automated driving—Environmental footprint effects of vacant trip time, Journal of Japan Society of Civil Engineers Ser D3, 74(5), I_889–I_896, https://doi.org/10.2208/jscejipm.74.I_889 (In Japanese).
11. Kii, M., Yokota, A., Gao, Z., and Nakamura, K. (2017), The effect of urban conditions on dissemination of shared fully-automated vehicles, Journal of Japan Society of Civil Engineers Ser D3, 73(5), I_507–I_515, https://doi.org/10.2208/jscejipm.73.I_507 (In Japanese).
12. Kilde, N. A., and Larsen, H. V. (2001), Scrapping of passenger cars in EU-15 in the years 1970 to 2015, Risø National Laboratory, Systems Analysis Department, Risø-I-1728 (EN), https://www.researchgate.net/profile/Helge-Larsen/publication/257536624_Scrapping_of_passenger_cars_in_EU-15_in_the_years_1970-2015/links/5a27ec75aca2727dd883e04d/Scrapping-of-passenger-cars-in-EU-15-in-the-years-1970-2015.pdf (accessed on July 31, 2022).
13. Latimer, T., and Meier, P. (2017), Use of the experience curve to understand economics for at-scale EGS projects, In: The 42nd workshop on geothermal reservoir engineering, US: California, https://pangea.stanford.edu/ERE/pdf/IGAstandard/SGW/2017/Latimer.pdf (accessed on July 31, 2022).
14. Lee, S., Yang, C., and Lee, E. (2015), ICT product diffusion in US and Korean markets, Industrial Management & Data Systems, 115(2), pp. 270–283, https://doi.org/10.1108/IMDS-09-2014-0254.
15. Litman, T. (2022), Autonomous vehicle implementation predictions: Implications for transport planning, Victoria Transport Policy Institute, http://www.vtpi.org/avip.pdf (accessed on April 17, 2022).
16. Mazur, C., Offer, G. J., Contestabile, M., and Brandon, N. (2018), Comparing the effects of vehicle automation, policy-making and changed user preferences on the uptake of electric cars and emissions from transport, Sustainability, 10(3), pp. 1–19, https://doi.org/10.3390/su10030676.

17. Miyoshi, H., Chapter 8: Jidō unten no shakai heno Inpakuto [Social impact of automated driving], in Satake, M., Iida, Y., and Yanagawa, T. (eds.) *Abe no Mikusu no Seihi* [*Success or Failure of Abenomics*], pp.142–162, Japan Economic Policy Association (JEPA), Keiso Shobō (In Japanese), 2019.

18. National Institute of Population and Social Security Research (IPSS) (2018), Regional Population Projections for Japan: 2015–2045, https://www.ipss.go.jp/pp-shicyoson/e/shicyoson18/t-page.asp (accessed on July 31,, 2022).

19. Nihon Bus Association (2020), Nohon no Basu Jjgyō 2020 [Bus Business in Japan 2020], 59, https://www.bus.or.jp/about/pdf/2020_busjigyo.pdf (In Japanese, accessed on October 21, 2022).

20. Nykvist, B., and Nilsson, M. (2015), Rapidly falling costs of battery packs for electric vehicles, Nature Climate Change, 5(4), pp. 329–332. https://doi.org/10.1038/nclimate2564.

21. Parvin, A. J., and Beruvides, M. G. (2021), Macro patterns and trends of U.S. consumer technological innovation diffusion rates, Systems, 9(1), 16, https://doi.org/10.3390/systems9010016.

22. Rogers, E. M. (2003), Diffusion of Innovations, 5th ed., New York, The Free Press.

23. The Society of Automotive Engineers (SAE) (2021), SAE J3016_2014_202104 Recommended Practice: Taxonomy and Definitions for Terms Related to Driving Automation Systems for On-Road Motor Vehicles, revised 2021–04–30, https://www.sae.org/standards/content/j3016_202104 (accessed on October 21, 2022).

24. Schmidt, O., Hawkes, A., Gambhir, A., and Staffell, I. (2017), The future cost of electrical energy storage based on experience rates, Nature Energy, 2, Article 17110, https://doi.org/10.1038/nenergy.2017.110.

25. Statistics Bureau, Ministry of Internal Affairs and Communications (MIC) (2015), Population Census 2015, https://www.stat.go.jp/english/data/kokusei/index.htm (accessed on July 31, 2022).

26. Suda, Y., and Miyoshi, H. (2021), Development of Assessment Methodology for Socioeconomic Impacts of Automated Driving Including Traffic Accident Reduction, Cross-ministerial Strategic Innovation Promotion Program (SIP), SIP 2nd Phase: Automated Driving for Universal Services-Mid-Term Results Report (2018–2020), pp.136–141 https://en.sip-adus.go.jp/rd/rd_page03.php (accessed on October 21, 2022).

27. Trommer, S., Kolarova, V., Fraedrich, E., Kröger, L., Kickhöfer, B., Kuhnimhof, T., Lenz, B., and Phleps, P. (2016), Autonomous Driving: The Impact of Vehicle Automation on Mobility Behaviour, Institute for Mobility Research, https://elib.dlr.de/110337/ (accessed on July 31, 2022).

Overall Comparison Between Germany and Japan in Relation to Social Impact of Connected and Automated Driving

Oguchi Takashi, Suzuki Shoichi, Christine Eisenmann, and Torsten Fleischer

Abstract The chapter "Setting the Scene for Automated Mobility: A Comparative Introduction to the Mobility Systems in Germany and Japan" provides an overview of the framework conditions characterizing the transport systems in Germany and Japan. Following two chapters investigate relation of the governance style to regulatory changes and resource allocation in the new technology development in Japan, and a business analysis and prognosis of the ride hailing market in Germany. The chapter 5 discusses the social acceptance of CAD based on the common survey conducted both in Japan and Germany. Following two chapters include the investigations on various parameters of the German transport system applying three consecutive transport models and scenario analyses, and the effects of the diffusion of CAD on the Japanese transport system applying a model developed by the authors in charge. Based on the discussions above, this chapter discuses several points relating to public expectations, touching on the similarities and differences between Germany and Japan; such as public acceptance affecting CAD diffusion, common expectations of groups and individuals, attitude of car industry, mobility services expected to be realized, differences in forms of residence, decision-making toward diffusion, expectations for the type of CAD initially introduced, forming a correct understanding of CAD by citizens, and risks that may affect expectations.

O. Takashi (✉) · S. Shoichi
The University of Tokyo, Tokyo, Japan
e-mail: takog@iis.u-tokyo.ac.jp

S. Shoichi
e-mail: suzuki41@iis.u-tokyo.ac.jp

C. Eisenmann
German Aerospace Center (DLR), Berlin, Germany

T. Fleischer
Karlsruhe Institute of Technology, Karlsruhe, Germany
e-mail: torsten.fleischer@kit.edu

This volume presents the main results of a bilateral research collaboration on Connected and Automated Driving (CAD) between Germany and Japan, initiated and supported by the German Federal Ministry of Education and Research (BMBF) and the Cabinet Office (CAO) of the Japanese Goverment. Researchers from both countries discussed and examined potential future impacts of CAD on the transport system, resulting in the six chapters discussed below.

Chapter "Setting the Scene for Automated Mobility: A Comparative Introduction to the Mobility Systems in Germany and Japan" provides a broad overview of the framework conditions that characterize the transport systems in Germany and Japan today (2022). It shows that while Japan and Germany are in some ways very different, there are also important similarities, and that these two countries are pertinent comparative case studies for CAD as they represent the prototypical path for mobility development. This chapter lays the groundwork for understanding the context for the introduction of vehicle automation, and provides a framework for interpreting the findings of the subsequent chapters in the overall context of the two national transport systems.

One specific framework condition that relates to the development and deployment of automated driving technologies and services in Japan and Germany are policy processes. Chapter "Governance, Policy and Regulation in the Field of Automated Driving: A Focus on Japan and Germany" investigates in particular how the governance style relates to regulatory changes and resource allocation in the development of new technologies and innovations in the respective society.

Chapter "Business Analysis and Prognosis Regarding the Shared Autonomous Vehicle Market in Germany" analyzes the current shared vehicle service market in Germany, and applies a holistic business analysis to derive future scenarios as to how this ride-hailing market may evolve with the introduction of shared autonomous vehicles (SAVs). The results show that, due to higher utilization rates, if the SAV diffusion rate was 100%, then around one third of the current shared vehicle fleet in Germany would be needed to serve today's demand. However, although a projected SAV service customer price level range of about 0.6 EUR is significantly lower than current cost of use for ride-hailing services, this may not be low enough to convince a considerable number of passengers to switch from their privately-owned car to SAV services.

Chapter "Social Acceptance of CAD in Japan and Germany: Conceptual Issues and Empirical Insights" sets out how acceptance of CAD is key to its diffusion, and looks at ways to conceptually capture and define social acceptance and understand the full scope of the concept. The analysis is enriched with empirical insights based primarily on quantitative research conducted in Japan and Germany. The chapter discusses changes in attitudes towards CAD and explores how automated vehicles (AVs) have been covered in newspaper reporting. It concludes that in order to achieve a "soft landing" for AVs in society, discussions that involve industry, government, academia, the private sector, and citizens will be essential, and further comparisons with countries that have different societal and cultural backgrounds will be important contributions to the discussion.

Chapters "Transportation Effects of Connected and Automated Driving in Germany" and "Transportation Effects of CAD in Japan" address the diffusion of CAD in Germany and Japan. Both analyses assume a wide deployment of CAD in the late 2030s and show model results for the year 2050. Due to different modelling suits and projection methods, the analyses are shown in two distinct chapters. Three consecutive transport models and scenario analyses are used in chapter "Transportation Effects of Connected and Automated Driving in Germany" for the CAD diffusion analyses for Germany. The results show that while privately-owned AVs are likely to increase car density, shared services might reduce the number of cars in 2050. However, the effects are rather limited, with the highest decrease in car density of −4% in a scenario of automated services in both urban and rural areas, and a share of privately-owned AVs of 43–44% in the 2050 passenger car stock. This mixed traffic of automated and conventional vehicles implies that AVs will not be able to realize their full potential regarding safety and efficiency. It also indicates that despite decreasing car density through shared services in urban areas, new transportation modes compensate for this effect by introducing new vehicles. This might attract trips from all other existing means of transportation, thus resulting in an increase of vehicle-kilometers-traveled of about 5%, which might lead to further congestion. The authors argue that in order to cope with the environmental implications, it is necessary to provide a political framework which stresses the advantages of CAD. Chapter "Transportation Effects of CAD in Japan" addresses the effects of the diffusion of CAD on the Japanese transport system by means of a modelling approach. The results show that higher expectations are more effective than lower costs in promoting adoption of the novel transportation paradigm of driverless taxis. This indicates that an appropriate enhancement of consumer expectations will be crucial for ensuring the spread of AVs. The results also show that enhanced expectations regarding AVs are likely to increase total automobile travel distances by inducing a modal shift toward driverless taxis. This may lead to higher energy consumption and more traffic congestion, as well as increased demand for government spending in areas such as road infrastructure.

This concluding chapter presents the issues identified in chapters "Setting the Scene for Automated Mobility: A Comparative Introduction to the Mobility Systems in Germany and Japan–Transportation Effects of CAD in Japan", and in discussions at the joint workshops held in Kyoto in May and October 2022, which researchers from both countries consider to be particularly significant, and identifies those which should be tackled through further joint research activities in close partnership between Japan and Germany.

1 Key Issues Raised Through the Research Collaboration

The following sections discuss the key issues raised through the joint research activities.

1.1 Social Expectations

Social acceptance of CAD will affect its diffusion. To discuss this point, it is necessary to clarify whether the focus is on the expectations of the individuals respectively the expectations and attitudes of a group of individuals, including experts in different fields, that lead their activity coordination. The former is discussed in chapter "Social Acceptance of CAD in Japan and Germany: Conceptual Issues and Empirical Insights", and the latter in chapters "Governance, Policy and Regulation in the Field of Automated Driving: A Focus on Japan and Germany" and "Business Analysis and Prognosis Regarding the Shared Autonomous Vehicle Market in Germany". Chapters "Transportation Effects of Connected and Automated Driving in Germany" and "Transportation Effects of CAD in Japan" include both perspectives.

Each stakeholder should carefully endeavour to understand whether these discussions focus on the individual expectations or group expectations. These two focus points are intended for consideration by Original Equipment Manufacturers (OEM) and governmental activity and inter-organisational coordination, when considering the user expectations of CAD. This expectation can be interpreted as one of the adjustment mechanisms. In fact, the spread of advanced technologies such as CAD requires not only the acceptance of citizens, but also the actions of governments and industries such as OEMs to create a legal framework, develop and sell products, and obtain support from the relevant authorities.

In addition, when discussing the expectations of society, an understanding of the limitations of our knowledge concerning the social learning process that will lead to the diffusion of CAD is necessary. Social learning plays a role in the dissemination of complex technologies in various ways. Expectations and attitudes within the general public may change as people become more accustomed to them. Innovators may change product specifications or deployment strategies after experiencing public reactions to new technologies. Regulators try to balance the interests of different groups and to adapt regulatory frameworks as they learn more about the qualities and limitations of different technologies, as well as about the intended and unintended consequences and implications of their broader use. We do not know when—and exactly how—this will happen; however, we do know that social learning will influence CAD diffusion. It can promote it if adequately understood and organized, or it can hamper it if ignored or mismanaged. Decision-makers need to understand that these processes will evolve in the coming decades.

Before the COVID-19 pandemic, there were many expectations that the new mobility provided by CAD would solve numerous social problems and create new business. However, much has changed compared to the situation before the advent of the COVID-19 pandemic in 2020.0, and the attitude of the car industry and OEMs seems to have withdrawn. As described in chapter "Business Analysis and Prognosis Regarding the Shared Autonomous Vehicle Market in Germany", a number of OEMs had difficulties to determine their own business models to promote CAD, and stopped or slowed down promoting service-oriented business models. To this day, obtaining CAD services or vehicles from major OEMs is difficult in both Germany and Japan, even if local governments are seeking new public transport services. A limited number of companies offer CAD vehicles, with limited levels of technology.

1.2 Commonalities and Differences Between Japan and Germany

In both Japan and Germany, the car industry is a major industry in the national economic system as illustrated in chapter "Setting the Scene for Automated Mobility: A Comparative Introduction to the Mobility Systems in Germany and Japan". But both German and Japanese experts expect Level 4 automated driving of privately-owned cars to be difficult to realize for the time being.

In both Germany and Japan, there is a common understanding that there are two versions of diffusion paths for CAD. One path is the expansion of the current business model of the car industry that promotes personally-owned automated cars with increased safety, higher comfort and increased mobility options, in order to ensure that car owners appreciate the value of the products, thus leading to the diffusion of personally-owned CAD cars. The other path employs CAD as a new mobility service for citizens in both rural and urban environments, and contributes to the provision of minimum mobility service levels and to the reduction of inefficient vehicle travel to support climate change challenge management. Figure 1[1] illustrates these two paths, and shows that automated driving in mobility services, where the Operational Design Domain (ODD) in which automated driving system works can be limited, can achieve Level 4 at an early stage, while automated driving of privately-owned cars, where a wider ODD must be set, will need to wait for technology to evolve.

To ensure that the existing automotive industry successfully realizes the transformation to CAD, two paths of CAD evolution are conceivable. The first path involving privately-owned passenger vehicles is supported by stakeholder groups that prefer to foster support for an improved level of automation, and supply financial support to the existing industry. These stakeholder groups are strong in Germany because of the extensive OEMs. This attitude is thought to be brought about by the belief that the automotive industry supports the countries economy and should be supported by the government. The second process is supported by stakeholder groups that explore ways to transform mobility systems and make them sustainable, more collective, and less energy-demanding. The groups supporting these two different evolution paths compete with each other.

Figure 1 suggests that the required level of automation may vary for different types of mobility service. For privately-owned cars, the expectation is that the automation level should gradually be raised from Level 2 to Level 3, approaching Level 4. Conversely, for mobility services such as public transport, Level 2 is not supported by

---

[1] These two paths were originally proposed in a SIP-adus project in 2016, and also reported in the 4th SIP-adus workshop in November 2017.

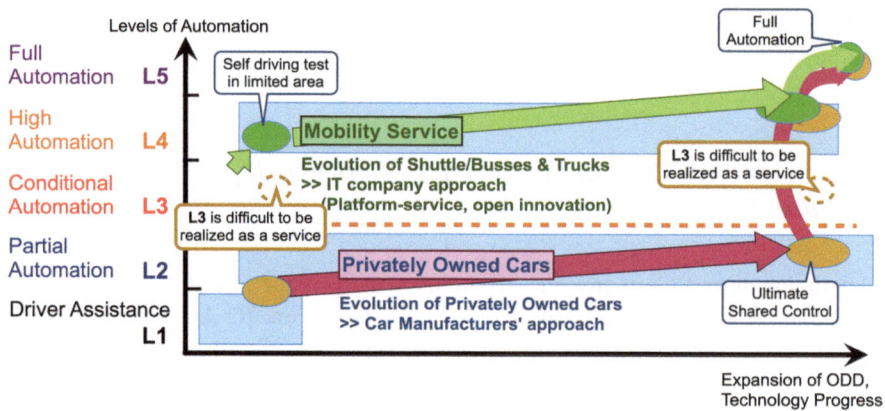

Fig. 1 Overview of the two paths of CAD evolution (modified from Sakai et al. [1])

any stakeholder groups. Although a Level 3 mobility service requires a transitional period, an expectation exists that Level 4 (without a driver), will be significantly appreciated because in addition to the operational flexibility it provides the possibility of considerably reducing the labour costs of human drivers, should the condition of a carefully selected and confirmed ODD be met.

The ODD for the two types of mobility differs completely. Technological, political, and societal expectations differ for different mobility services. However, it is clearly observed that the discussions in the public domain include both types of mobility. As the expectations and assumptions of individuals could differ in discussions or considerations with an unclear definition of mobility, 'mobility' should be defined in advance so that the acceptance or expectations of society can be discussed.

Regarding mobility services expected to be realized, in Japan there are high expectations that CAD will help provide possible mobility to elderly people in rural areas. Although automation itself is not necessary from a user point of view, there is a high expectation of CAD from local authorities and mobility service providers because it is believed that labor costs could be minimized and driver shortages alleviated by introducing CAD. In Japan, public transport services have been stopped in some rural areas, resulting in local governments starting to operate these services as welfare services. In Germany, local governments generally operate the mobility services. But in both Germany and Japan, although it is unnecessary to make a profit from public transport services, a serious concern is the reduction of costs.

In Germany, CAD is expected not only to reduce costs, but also to increase energy efficiency and reduce CO_2 emissions by ensuring a flexible operation of services. Energy conservation has been emphasized from both the environmental and economic perspectives since the oil crisis in the 1970s, and has gained high attention in the course of rising energy costs due to the Russian invasion of Ukraine in 2022.

Conversely, transportation operators and citizens in rural areas in Japan are not seriously considering the reduction of CO_2 emissions, although policymakers request stakeholders to implement steps that may lead to carbon neutrality. In addition, in Japan, there is a high expectation that driverless CAD will solve the driver shortage issue; however, this is not the case in Germany.

Regarding demography and settlement patterns, Germany and Japan have somewhat different forms, as described in chapter "Setting the Scene for Automated Mobility: A Comparative Introduction to the Mobility Systems in Germany and Japan". In Germany, more people live in dispersed rural areas. However, in the rural areas of both Germany and Japan, mobility without a privately-owned car is hard to ensure. From a technical point of view, the introduction of CAD in rural areas is more straightforward where traffic volumes are relatively low, making a consensus for introducing mobility service by CAD easier to obtain. However, the funding resource shortage is a serious issue in most rural areas, except in some wealthy areas where factories of global manufacturers are located and local authorities can secure adequate tax revenue to retain mobility services. In Germany, politicians recognize that one major concern of rural populations s is the lack of options concerning public transport services, although this is not the case in Japan because basic consensus on public transport is different, as described in chapter "Setting the Scene for Automated Mobility: A Comparative Introduction to the Mobility Systems in Germany and Japan". Thus, the improvement of public transport can be politically motivated.

Regarding the expected type of CAD to be initially introduced, as described in chapter "Setting the Scene for Automated Mobility: A Comparative Introduction to the Mobility Systems in Germany and Japan", both Germany and Japan have made a major shift from rail to truck freight. The initial introduction and diffusion of CAD may begin with long-distance truck freight transport. Moreover, with the diffusion of CAD there is the expectation that the understanding and acceptance of society will increase. However, the discussion at the joint workshops in Kyoto reached a conclusion that in last- or first-mile delivery, where drivers are required to fulfil various roles, such as securing and management of loads, loading, and unloading, the introduction of CAD at an early stage from an economic viability point of view may be challenging in both Japan and Germany.

1.3   Decision-Making Toward Diffusion

As described in chapter "Transportation Effects of CAD in Japan", studies in Japan have shown that prior to the actual diffusion of CAD in society, improvement in CAD knowledge and expectations has a greater impact on its projected diffusion than the CAD cost requirement. However, these results should be carefully considered. Once citizens become acquainted with CAD, the cost of CAD will become one of the most important factors affecting its diffusion.

The diffusion of CAD has a complex mechanism. The price of a CAD vehicle could be evaluated by combining the cost with numerous other factors such as the safety level, ease of use of the CAD vehicle, applicability to personal travel needs in a given situation, personal preference between using a privately-owned car or public transport, and electrification levels. These complexities are disjointed and should not be oversimplified, so that an incorrect understanding is not offered by experts to policy makers.

1.4 Possible Risks Arising from the Diffusion of CAD

Discussions in the joint workshops in Kyoto, based on the results of the Chapters, addressed how in order to increase societal acceptance towards the diffusion of CAD, the following risks should be heeded. There are two key points: the use of digital infrastructure, and global supply chains at the manufacturing stage.

Since the realization of CAD is largely dependent on digital infrastructure, resilience against the risk of remote hacking is necessary. Recent geopolitical shifts may lead to the opportunity for citizens to witness digital infrastructure attacks and people may become reluctant to use services using digital infrastructure.

For the manufacture of high-end CAD products by Japanese and German OEMs, materials, semiconductors, parts, and so forth are supplied through the global supply chain. Considering the impact of COVID-19 and economic conflicts between major state powers, the reliance on a single country for any supply is extremely risky. In Japan, where such problems have arisen several times in the past, OEMs have already considered and implemented measures such as diversifying supply bases to Southeast Asian and other countries. Conversely, in Germany, discussions have only just begun on the necessity of dispersing the supply of energy, parts, materials, and so forth.

In addition, the simple replacement of privately-owned cars with CAD, especially in city centers, may lead to serious impacts such as a significant increase in vehicle-kilometers-traveled, as described in chapter "Transportation Effects of Connected and Automated Driving in Germany". Conversely, should citizens change their travel behavior from privately-owned cars to shared services provided by CAD, the total vehicle-kilometers-traveled would decrease. Given the risk that the total number of vehicle-kilometers-traveled will increase, policy efforts are needed to encourage and enable behavior change. It is necessary to clearly indicate the pros and cons of CAD to citizens and promote well-informed understanding of the expected impact for citizens and other stakeholders.

2 Future Avenues of Research

Through the collaborative activities, we were able to answer many of our research questions. At the same time, our joint research activities have led to the emergence of new research questions. These avenues for future research in the field of impact assessment are outlined below.

2.1 Social Acceptance and Adoption

In public and academic debates about the future of CAD, numerous stakeholders (policymakers, industry executives, scientists) stress the importance of "social acceptance" for the successful adoption and diffusion of AVs (chapter "Social Acceptance of CAD in Japan and Germany: Conceptual Issues and Empirical Insights"). This is both encouraging and challenging.

It is encouraging because at least implicitly, this acknowledges that introducing a new mobility technology is also a social program whose outcomes depend on the interplay of three elements:

- the features of the new technology and the mobility services it is intended to enable or improve,
- the regulatory framework (both "hard", i.e. legal, and "soft", i.e. institutional), and
- the attitudes and actions of users and non-users alike.

How these elements and their interdependences are methodologically captured and studied, and how the knowledge gathered by their investigation is translated into action by the variety of innovation actors that take part in CAD development and deployment, is decisive for the actual innovation and diffusion pathways of CAD. It also plays an important role in how these elements contribute to achieving the promises of CAD, as well as to wider societal goals for future mobility systems.

At the same time, the importance and complexity of "social acceptance" in an innovation network create substantial challenges for communication and cooperation between innovation actors, especially across different domains. Within science, there is so far no broadly-shared definition of "acceptance" in general, or "social acceptance" in particular. In substantial parts of the scientific literature, under the umbrella term of "acceptance", at least three phenomena are investigated that could be more precisely described as three different foci of "user acceptance" or "citizens' acceptance":

- the perceptions of and attitudes towards certain products or services,
- the (stated) intention to use them, and
- their actual use.

These are obviously interrelated, but it is well-known that they do not simply translate into each other. Innovation actors in other fields, e.g., in environmental technologies or sustainable consumer products, have experienced remarkable "mismatches" between attitudes and actual use. Since CAD has not yet been introduced at a scale where it has measurable influence on everyday mobility behavior, empirical evidence about changes in usage patterns is still missing. But one might reasonably assume that similar differences or "mismatches" will also occur in this field. It is, e.g., currently still open how attitudes towards autonomous vehicles among interested citizens (the subpopulation usually interviewed during field trials) relate to the actual usage of a new AV-based mobility service within a certain area (which would define the actual impacts). This creates some uncertainty for diffusion scenarios and policy decision-making since the input data so far mainly depend on information derived from early attitude studies.

We recommend that these limitations should be communicated more explicitly among stakeholders and to the wider public, and that the scenarios are adapted as soon as new information becomes available. In science-industry-policymaking debates, the term "social acceptance" appears to have a slightly different meaning than in the academic discussion outlined above. We read it as a metaphor for approaches which are trying to capture the societal dynamics (including diffusion, but also, i.e., user resistance or policy conflict) related to CAD. First studies into certain aspects of this subject have been published, but the large number of innovation actors involved in CAD development and deployment, as well as the vast legal and institutional framework which influences the field, create a complex innovation landscape whose elements and interactions are not yet sufficiently well-captured. Further research should be dedicated to this problem.

2.2 Diversity of Expectations

It can be observed that different stakeholders have different expectations about how CAD can contribute to solving the current problems in the transport system. Some of the narratives, expectations and illusions are unrealistic and exaggerated when reflecting them in serious research studies, such as the studies conducted in chapters "Business Analysis and Prognosis Regarding the Shared Autonomous Vehicle Market in Germany", "Social Acceptance of CAD in Japan and Germany: Conceptual Issues and Empirical Insights", "Transportation Effects of Connected and Automated Driving in Germany" and "Transportation Effects of CAD in Japan". Transport researchers agree that AVs won't be available soon for solving today's transport problems.

Research needs to systematically compile and scrutinize the motivations of the different stakeholders for drawing their narratives. During the joint workshops in Kyoto it was discussed that, for example, providers of AVs and automated services want to convince venture capitalists of their company's merits; therefore, positive expectations towards the market potential and profit generation are key. Moreover, it needs to be researched whether and to what extent the—sometimes exaggerated— narratives of the different stakeholders shape the expectations of society towards CAD. It was also argued during the workshops that the risks of these interrelationships leading to unexpected feedback loops and undesired outcomes should be explored. For example, young people's choice of occupation may be influenced by the expectation that automation will lead to an early rationalization of occupational profiles (e.g., truck drivers, bus drivers). Career choices based on optimistically exaggerated expectations may then lead to an unintended workforce shortage of these occupational profiles in the near future.

2.3 Impact Estimation in Combination with Other Factors in the Transport Sector

Existing research on the impact of vehicle automation on travel and transport has mostly looked at the isolated influence of vehicle automation. This means that—as in the research reported in this book—models were utilized to analyze the differences between business-as-usual-scenarios and business-as-usual-plus-vehicle-automation scenarios. Our findings suggest that vehicle automation makes a difference, but does not lead to a fundamentally different transport ecosystem. The impacts found are not disruptive and make a limited contribution to getting closer to climate- or broader sustainability goals.

However, the interaction between vehicle automation and other factors or measures influencing the transport sector have not yet been explored sufficiently. It may well be the case that vehicle automation unfolds a different potential if combined with—for example—fundamentally altered parking or public transport policies. This is because vehicle automation and other factors may synergetically reinforce—or counterbalance—each other.

A starting point for exploring the potential of vehicle automation in combination with other factors is the formulation of desirable mobility futures, i.e., normative scenarios of how societies envision their future. Formulating such desired scenarios as societal objectives is not a task primarily for academia. Instead, each society—be it on the municipal, national or even a wider geographic level—has its own mechanisms to develop such visions, e.g., through democratic or participatory processes. Hence, the research task in this context concerns approaches of how to integrate vehicle automation as a factor among many such processes, and devise methods to identify the contribution of vehicle automation in combination with other factors. Such approaches and methods may include a wide range of qualitative and quantitative techniques, ranging from scenario planning to travel-demand modeling.

One thing, however, seems clear: while the approaches to automation impact estimation so far were mostly expert systems, in the future it will be increasingly important to develop understandable and transparent methods that work well in co-productive processes which also involve non-expert stakeholders.

2.4 Wider Impacts of Vehicle Automation Beyond Transportation

The convergence of findings by different researchers on the impact of vehicle automation on transport in recent years indicates that we are increasingly able to obtain a better understanding of the likely developments induced by automation. Hence, analyses and models as presented in chapters "Business Analysis and Prognosis Regarding the Shared Autonomous Vehicle Market in Germany", "Transportation Effects of Connected and Automated Driving in Germany" and "Transportation

Effects of CAD in Japan" in this book were successful when it comes to the transport impact of automation. However, transport impacts are—in most cases—only a proxy for other benefits or changes. For example, if we take as an example increased trip length caused by decreased values of travel time: They essentially mean more freedom in destination choice, i.e. increased utility for travelers. However, more travel means more emissions and may—in the long term—also impact on land use and settlement structures.

Hence, there are wider consequences of vehicle automation which impact on many dimensions of sustainability, including the environment, safety, equity and the economy. Among these wider influences some stand out as specifically important but methodologically challenging. These include:

(a) land-use patterns and urban form as a consequence of residential and other location choice interacting with urban planning;
(b) equity impacts as caused—for example—by different capabilities to use automation technologies by different user groups;
(c) impacts of automation on logistics supply and value chains.

For these dimensions of the impact of vehicle automation, strong narratives and some conceptual exploration exist. However, quantitative analysis and modeling for these dimensions of vehicle automation impact is still in infancy and hence represents an important field for future research.

Particularly with regard to the equity issues, it is necessary to systematically investigate which political, institutional, infrastructural, planning and organizational frameworks are required for CAD to contribute to mobility justice. Important aspects include the aging of society, which is a key demographic issue for the future in both Japan and Germany.

Reference

1. Sakai et al. (2018): Study of the requirements for virtual traffic experimental environment for prior evaluation of traffic operation policy in expressways, Bimonthly Journal of Institute of Industrial Science, 70(2), 57–62, March 2018 (in Japanese). https://doi.org/10.11188/seisanken kyu.70.57.

Correction to: Acceptance and Diffusion of Connected and Automated Driving in Japan and Germany

Christine Eisenmann, Dennis Seibert, Torsten Fleischer, Ayako Taniguchi, and Takashi Oguchi

Correction to:
C. Eisenmann et al. (eds.), *Acceptance and Diffusion*
of Connected and Automated Driving in Japan and Germany,
https://doi.org/10.1007/978-3-031-59876-0

In the original version of this book, belated corrections for chapter "Setting the Scene for Automated Mobility: A Comparative Introduction to the Mobility Systems in Germany and Japan" have been incorporated. Additionally, Figures 5, 6, and 7 in chapter "Transportation Effects of CAD in Japan" have been revised for improved readability. The book has been updated with the changes.

The updated versions of these chapters can be found at
https://doi.org/10.1007/978-3-031-59876-0_2
https://doi.org/10.1007/978-3-031-59876-0_7

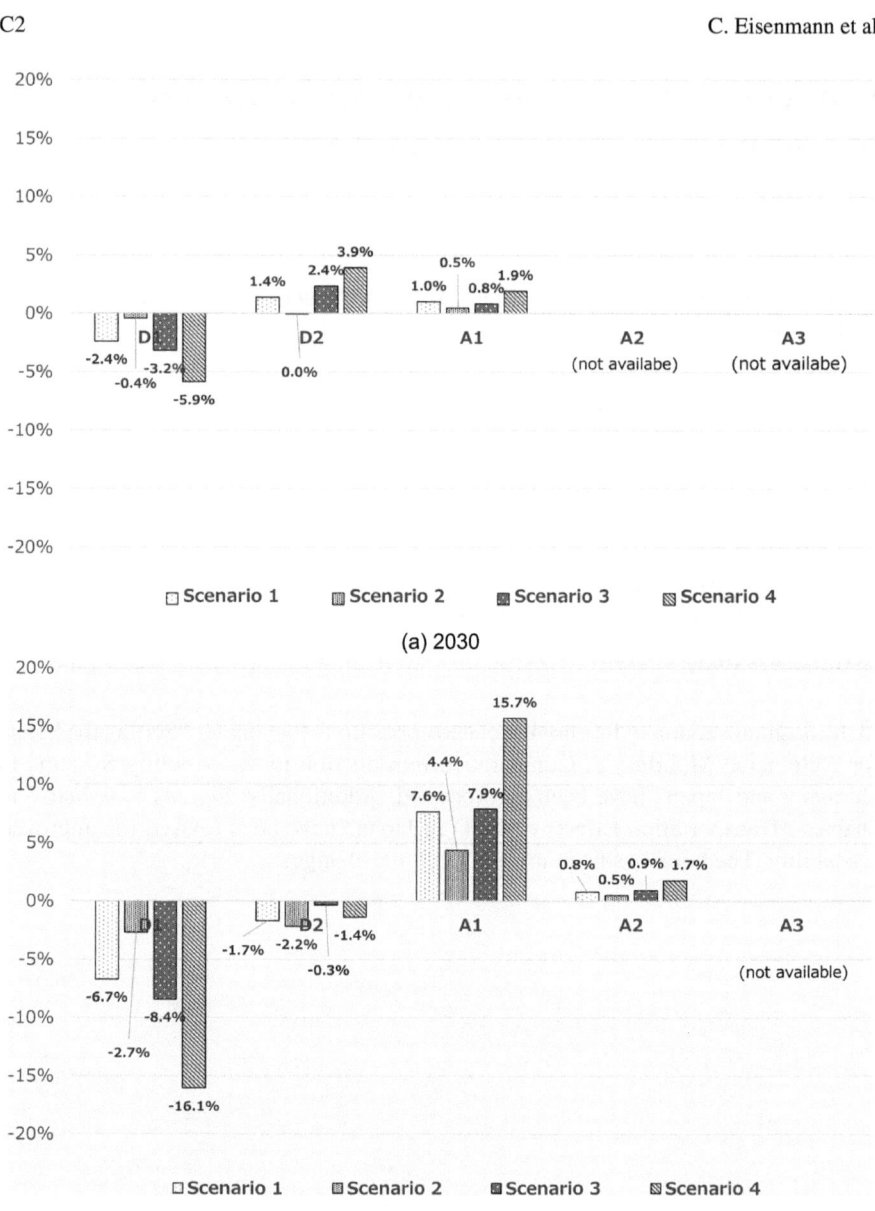

(a) 2030

(b) 2040

Fig. 5 Results of dynamic-model sensitivity analysis

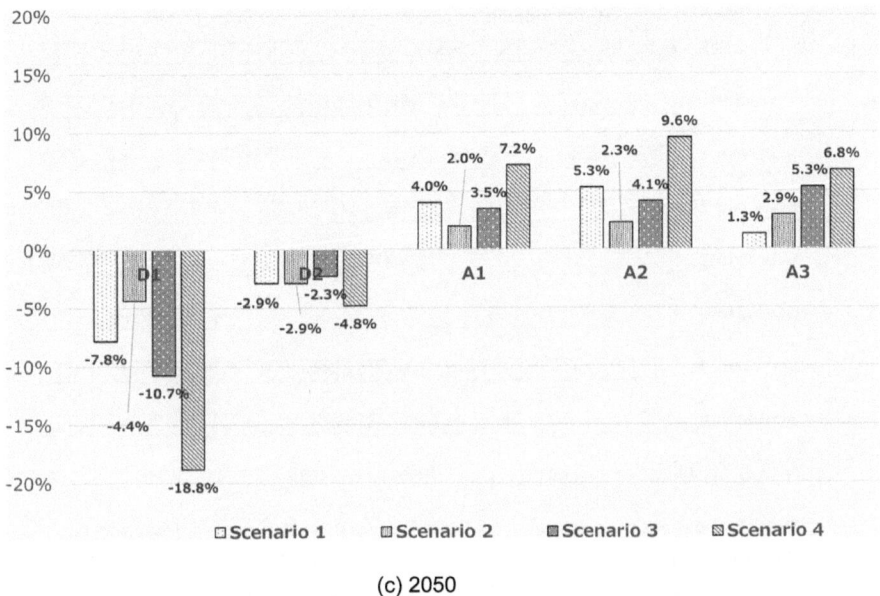

(c) 2050

Fig. 5 (continued)

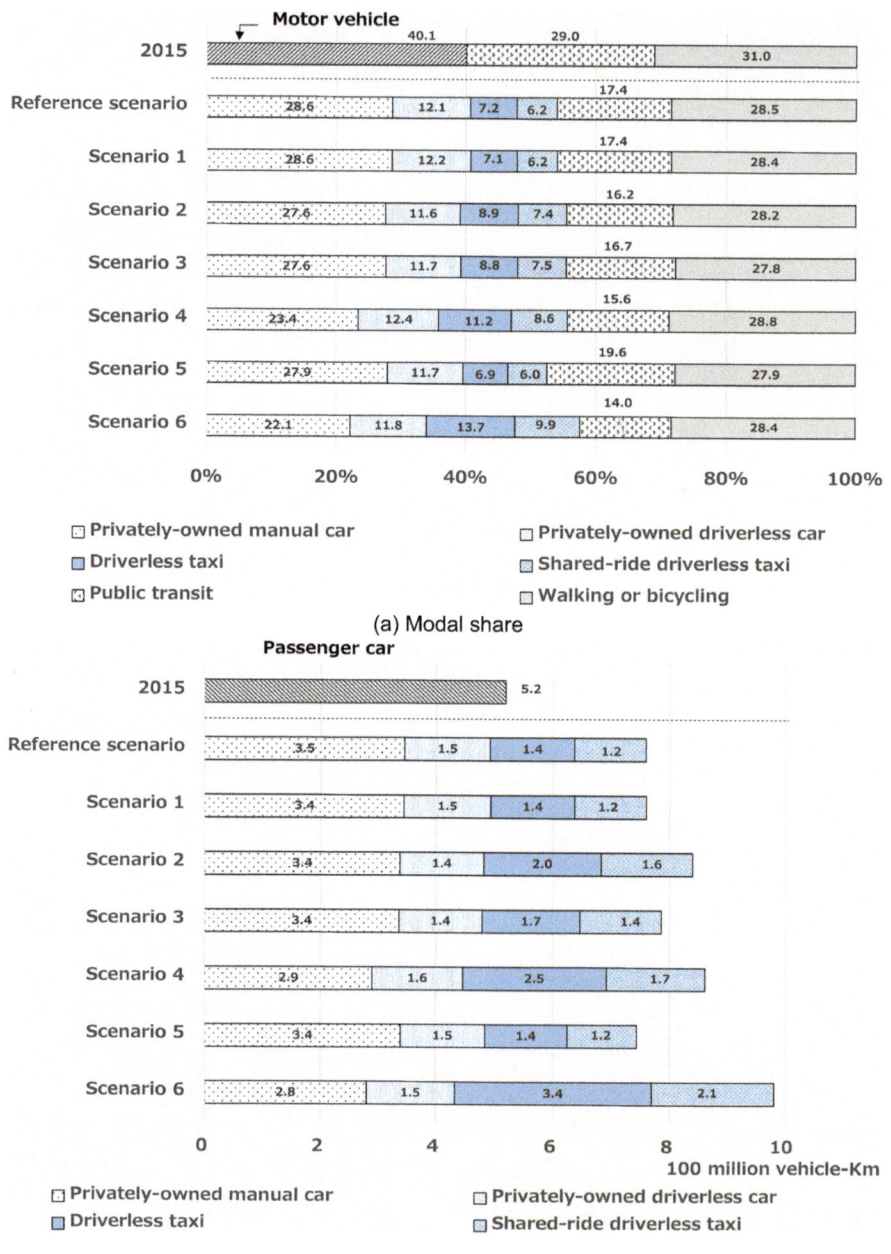

(a) Modal share

(b) Total daily travel distance

Fig. 6 Results of static-model simulations for geographical blocks containing the three major metropolitan areas Note: The 2015 values of (a) are calculated based on the data from the Nationwide Person Trip Survey provided by the Ministry of Land, Infrastructure, Transport and Tourism. The 2015 values of (b) and (c) are determined based on data from the Road Traffic Census provided by the Ministry of Land, Infrastructure, Transport and Tourism, and from the Automobile Inspection & Registration Association, respectively

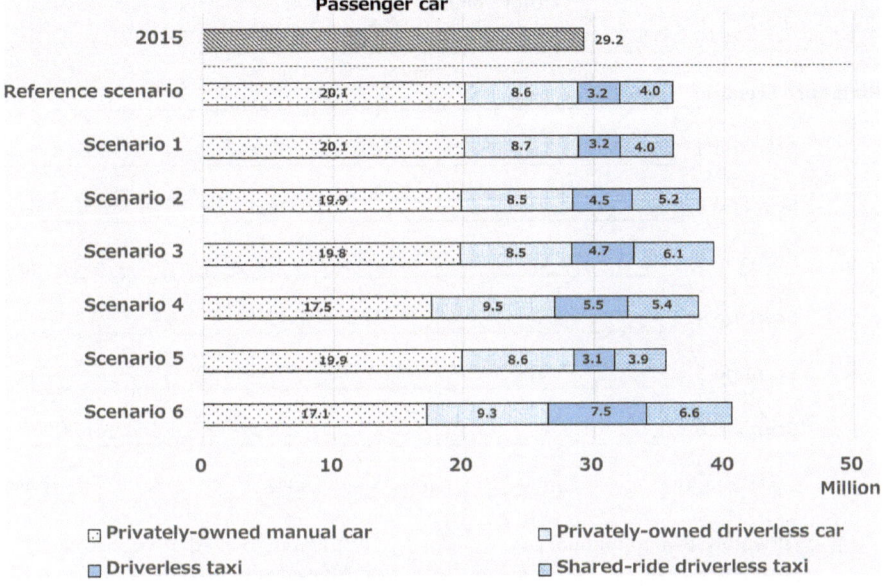

(c) Number of passenger cars owned

Fig. 6 (continued)

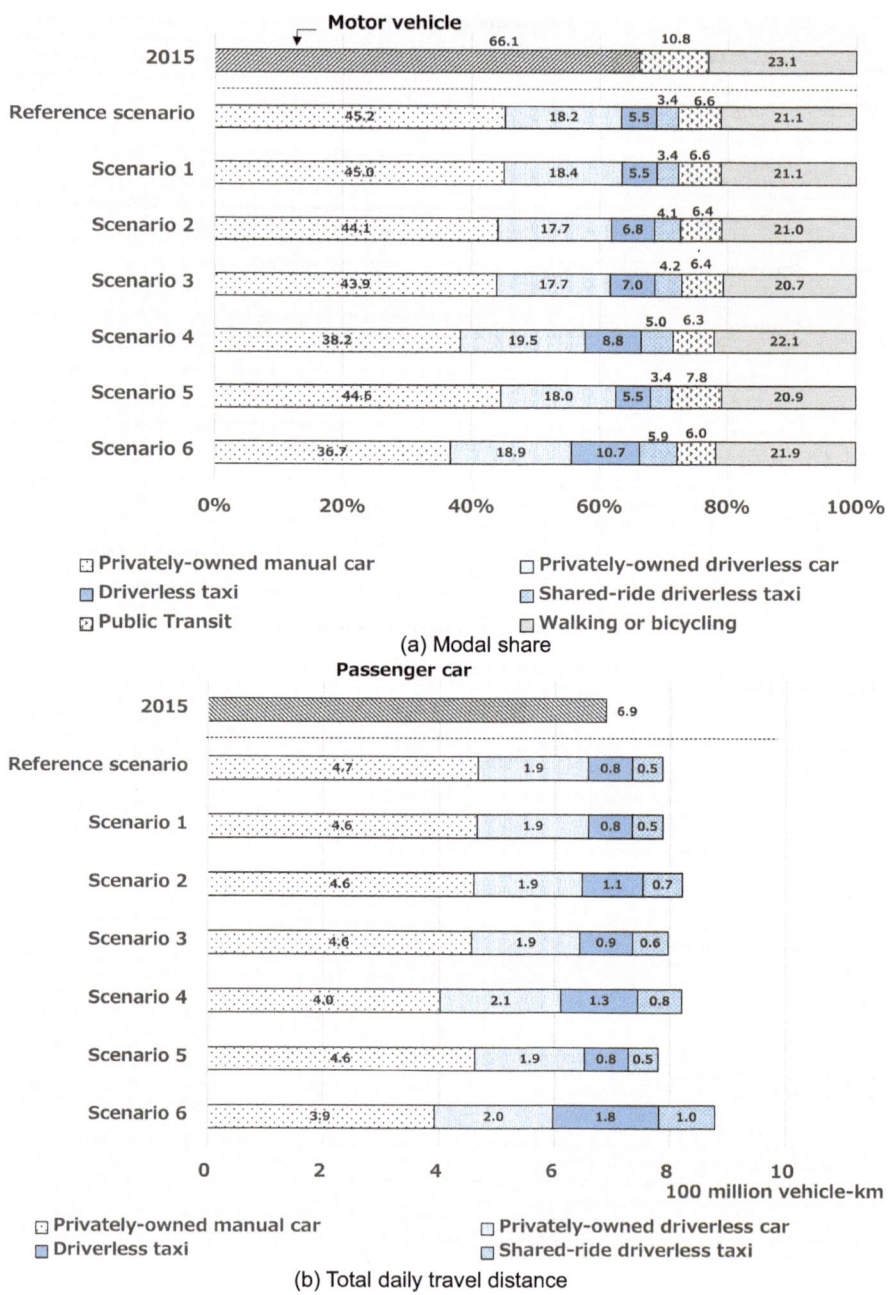

Fig. 7 Results of static-model simulations for geographical blocks not containing the three major metropolitan areas Note: The 2015 values of (a) are calculated based on the data from the Nationwide Person Trip Survey provided by the Ministry of Land, Infrastructure, Transport and Tourism.The 2015 values of (b) and (c) are determined based on data from the Road Traffic Census provided by theMinistry of Land,Infrastructure, Transport and Tourism, and from the Automobile Inspection & Registration Association, respectively

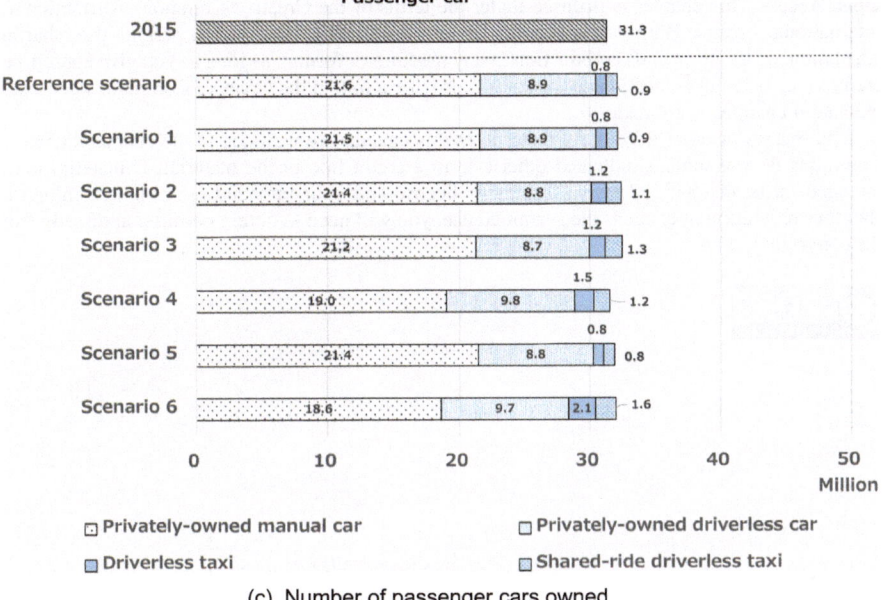

(c) Number of passenger cars owned

Fig. 7 (continued)